DESIGN CONSIDERATIONS FOR NUCLEAR POWER PLANTS AGAINST TSUNAMIS

The following States are Members of the International Atomic Energy Agency:

AFGHANISTAN
ALBANIA
ALGERIA
ANGOLA
ANTIGUA AND BARBUDA
ARGENTINA
ARMENIA
AUSTRALIA
AUSTRIA
AZERBAIJAN
BAHAMAS, THE
BAHRAIN
BANGLADESH
BARBADOS
BELARUS
BELGIUM
BELIZE
BENIN
BOLIVIA, PLURINATIONAL STATE OF
BOSNIA AND HERZEGOVINA
BOTSWANA
BRAZIL
BRUNEI DARUSSALAM
BULGARIA
BURKINA FASO
BURUNDI
CABO VERDE
CAMBODIA
CAMEROON
CANADA
CENTRAL AFRICAN REPUBLIC
CHAD
CHILE
CHINA
COLOMBIA
COMOROS
CONGO
COOK ISLANDS
COSTA RICA
CÔTE D'IVOIRE
CROATIA
CUBA
CYPRUS
CZECH REPUBLIC
DEMOCRATIC REPUBLIC OF THE CONGO
DENMARK
DJIBOUTI
DOMINICA
DOMINICAN REPUBLIC
ECUADOR
EGYPT
EL SALVADOR
ERITREA
ESTONIA
ESWATINI
ETHIOPIA
FIJI
FINLAND
FRANCE
GABON
GAMBIA, THE
GEORGIA
GERMANY
GHANA
GREECE
GRENADA
GUATEMALA
GUINEA
GUYANA
HAITI
HOLY SEE
HONDURAS
HUNGARY
ICELAND
INDIA
INDONESIA
IRAN, ISLAMIC REPUBLIC OF
IRAQ
IRELAND
ISRAEL
ITALY
JAMAICA
JAPAN
JORDAN
KAZAKHSTAN
KENYA
KOREA, REPUBLIC OF
KUWAIT
KYRGYZSTAN
LAO PEOPLE'S DEMOCRATIC REPUBLIC
LATVIA
LEBANON
LESOTHO
LIBERIA
LIBYA
LIECHTENSTEIN
LITHUANIA
LUXEMBOURG
MADAGASCAR
MALAWI
MALAYSIA
MALI
MALTA
MARSHALL ISLANDS
MAURITANIA
MAURITIUS
MEXICO
MONACO
MONGOLIA
MONTENEGRO
MOROCCO
MOZAMBIQUE
MYANMAR
NAMIBIA
NEPAL
NETHERLANDS, KINGDOM OF THE
NEW ZEALAND
NICARAGUA
NIGER
NIGERIA
NORTH MACEDONIA
NORWAY
OMAN
PAKISTAN
PALAU
PANAMA
PAPUA NEW GUINEA
PARAGUAY
PERU
PHILIPPINES
POLAND
PORTUGAL
QATAR
REPUBLIC OF MOLDOVA
ROMANIA
RUSSIAN FEDERATION
RWANDA
SAINT KITTS AND NEVIS
SAINT LUCIA
SAINT VINCENT AND THE GRENADINES
SAMOA
SAN MARINO
SAUDI ARABIA
SENEGAL
SERBIA
SEYCHELLES
SIERRA LEONE
SINGAPORE
SLOVAKIA
SLOVENIA
SOMALIA
SOUTH AFRICA
SPAIN
SRI LANKA
SUDAN
SWEDEN
SWITZERLAND
SYRIAN ARAB REPUBLIC
TAJIKISTAN
THAILAND
TOGO
TONGA
TRINIDAD AND TOBAGO
TUNISIA
TÜRKİYE
TURKMENISTAN
UGANDA
UKRAINE
UNITED ARAB EMIRATES
UNITED KINGDOM OF GREAT BRITAIN AND NORTHERN IRELAND
UNITED REPUBLIC OF TANZANIA
UNITED STATES OF AMERICA
URUGUAY
UZBEKISTAN
VANUATU
VENEZUELA, BOLIVARIAN REPUBLIC OF
VIET NAM
YEMEN
ZAMBIA
ZIMBABWE

The Agency's Statute was approved on 23 October 1956 by the Conference on the Statute of the IAEA held at United Nations Headquarters, New York; it entered into force on 29 July 1957. The Headquarters of the Agency are situated in Vienna. Its principal objective is "to accelerate and enlarge the contribution of atomic energy to peace, health and prosperity throughout the world".

SAFETY REPORTS SERIES No. 128

DESIGN CONSIDERATIONS FOR NUCLEAR POWER PLANTS AGAINST TSUNAMIS

INTERNATIONAL ATOMIC ENERGY AGENCY
VIENNA, 2025

Marketing and Sales Unit, Publishing Section
International Atomic Energy Agency
Vienna International Centre
PO Box 100
1400 Vienna, Austria
fax: +43 1 26007 22529
tel.: +43 1 2600 22417
email: sales.publications@iaea.org
www.iaea.org/publications

Printed by the IAEA in Austria
June 2025
STI/PUB/2103
https://doi.org/10.61092/iaea.i9ln-icob

IAEA Library Cataloguing in Publication Data

Names: International Atomic Energy Agency.
Title: Design considerations for nuclear power plants against tsunamis / International Atomic Energy Agency.
Description: Vienna : International Atomic Energy Agency, 2025. | Series: IAEA safety reports series, ISSN 1020–6450 ; no. 128 | Includes bibliographical references.
Identifiers: IAEAL 25-01757 | ISBN 978-92-0-101125-1 (paperback : alk. paper) | ISBN 978-92-0-101225-8 (pdf) | ISBN 978-92-0-101325-5 (epub)
Subjects: LCSH: Nuclear power plants — Design and construction. | Nuclear power plants — Safety measures. | Tsunamis. | Natural disasters.
Classification: UDC 621.039.58 | STI/PUB/2103

FOREWORD

Since the tsunami that was caused by the 2011 Great East Japan Earthquake, the importance of protecting nuclear installations against coastal flooding has received renewed attention. That event highlighted the need for deeper consideration of the design against tsunamis and the safety assessment of nuclear installations in relation to such hazards, including the potential benefits of a risk informed decision making framework. As a result, the engineering community has proposed a combined approach to the design of nuclear installations against tsunami scenarios that is based on both deterministic and probabilistic methods. IAEA guidance on such an approach could enhance current practice and the safety of nuclear sites that may be affected by a tsunami.

IAEA Safety Standards Series Nos SSG-18, Meteorological and Hydrological Hazards in Site Evaluation for Nuclear Installations, and SSG-68, Design of Nuclear Installations Against External Events Excluding Earthquakes, provide high level guidance on the development of hydrological and meteorological hazards, as well as recommendations for the design and safety assessment of nuclear installations. This publication complements those Specific Safety Guides and provides important insights into the various impacts of tsunamis regarding up to date and well validated engineering approaches to the safety of coastal nuclear installations. This publication is expected to be of value to designers and regulatory bodies.

The IAEA is grateful to all those who contributed to the drafting and review of this publication. The IAEA officer responsible for this publication was K. Nagasawa of the Division of Nuclear Installation Safety.

EDITORIAL NOTE

CONTENTS

1. INTRODUCTION

1.1. BACKGROUND

In the aftermath of the accident at the Fukushima Daiichi nuclear power plant (NPP) in 2011, significant attention was directed towards re-evaluating and enhancing tsunami design protocols for nuclear installations. The accident underscored the critical importance of anticipating and mitigating the impact of tsunamis on nuclear installations.

The events at the Fukushima Daiichi plant prompted a global reassessment of nuclear safety standards and the need for proactive measures to address potential vulnerabilities. The existing tsunami design, predicated on a design basis tsunami, came under scrutiny, as it was deemed insufficient to withstand the unprecedented scale of the 2011 tsunami. This involved considering scenarios beyond the previously defined design basis tsunami and incorporating lessons learned from the Fukushima Daiichi accident into related guidelines.

Furthermore, various types of impact and damage were observed not only at NPPs, but also extensively across the Pacific coast of Japan, during the tsunami. When assessing tsunami impacts, it is crucial to consider the maximum water level, as well as factors such as the hydrodynamic effects and impacts from waterborne missiles. Additionally, attention needs to be paid to the tsunami induced fires observed in Kesennuma, Japan. Depending on the layout of nuclear sites, it is necessary to install countermeasures and to take these impacts into account when designing nuclear installations, which was not a common practice before the Fukushima Daiichi accident.

Within this context, IAEA Safety Standards Series No. SSR-2/1 (Rev. 1), Safety of Nuclear Power Plants: Design [1], which includes a requirement on design extension conditions (DECs), was published in 2016. IAEA Safety Standards Series No. SSG-18, Meteorological and Hydrological Hazards in Site Evaluation for Nuclear Installations [2], which was published in 2011, is currently in the process of revision to integrate the most recent concepts, incorporating insights gained from the Fukushima Daiichi accident. Additionally, IAEA Safety Standards Series No. SSG-68, Design of Nuclear Installations Against External Events Excluding Earthquakes [3], released in 2021, provides comprehensive recommendations on tsunami design and evaluation for design basis tsunamis and beyond design basis tsunamis, encompassing diverse impacts of tsunamis.

1.2. OBJECTIVE

The main objective of this publication is to illustrate the diversity of impacts caused by tsunamis and to provide an overview of the design considerations for tsunami specific elements concerning design basis tsunamis and the evaluation of tsunami design to include beyond design basis tsunamis. The ultimate goal is to contribute to the enhancement of safety measures for nuclear installations in accordance with SSG-68 [3]. Guidance and recommendations provided here in relation to identified good practices represent expert opinion but are not made on the basis of a consensus of all Member States.

1.3. SCOPE

This publication addresses the elements of the design basis tsunami — including the design concept, the preliminary design and the final design — and their verification against the beyond design basis tsunami for nuclear installations.

Design activities need to be closely coordinated with the beyond design basis tsunami evaluations, incorporating methods such as tsunami margin assessments and tsunami probabilistic safety assessment activities. Furthermore, the process of designing nuclear installations to withstand tsunami effects necessitates a multidisciplinary approach involving expertise in the following areas:

— Nuclear safety;
— Seismic hazard (seismology, geology, geophysics, geotechnical engineering);
— Tsunami generation and propagation (oceanography, hydrology);
— Engineering disciplines (civil, structural, flooding, geotechnical, mechanical, electrical systems);
— Power plant operation.

1.4. STRUCTURE

This publication comprises the following sections. Section 2 provides an overview and introduction of the tsunami design process: the overall purpose of the tsunami design; potential effects of a tsunami on a nuclear installation; the overall approach; tsunami phenomena and loading conditions; and categorization of tsunami protection system elements. Section 3 defines the design basis tsunami and the beyond design basis tsunami loading conditions. Section 4 provides an overview of the design process for the tsunami protection system. Sections 5 and 6 discuss the design processes for a tsunami protection system comprising external and incorporated barriers, and mechanical and electrical equipment and distribution systems, respectively. Section 7 discusses design that addresses special issues of tsunamis. Section 8 discusses tsunami detection and warning systems. Section 9 presents the management system. Annexes I and II describe the practices followed in the United States of America and Japan, respectively, in relation to design against tsunami scenarios. These two countries are highlighted herein, as they have extensive standards and codes covering earthquakes and associated (concomitant) events, such as flooding associated with a tsunami.

2. MAIN APPROACH TO DESIGN AGAINST TSUNAMI EFFECTS

2.1. GENERAL CONSIDERATIONS IN TSUNAMI DESIGN

The objectives of tsunami design, as set out in IAEA Safety Standards and other Member State regulatory and guidance publications, are as follows. At the highest level, the objective is to ensure that NPPs are designed and operated safely, minimizing the radiation exposure to on-site and off-site personnel and the environment (see SSR-2/1 (Rev. 1) [1]). Some Member States quantify guidance on core damage frequencies and release frequencies. Examples include the following:

— United States of America: The goals of the United States Nuclear Regulatory Commission (NRC) are less than 10^{-4} per reactor-year for mean core damage frequency and less than 10^{-6} per reactor-year for mean large release frequency [4, 5].

— Canada: The Canadian Nuclear Safety Commission specifies three probabilistic performance goals: (a) small release frequencies of less than 10^{-5} per annum; (b) large release frequencies of less than 10^{-6} per annum; and (c) core damage frequencies of less than 10^{-5} per annum [6].

— Japan: The safety goals of the Nuclear Regulation Authority (NRA) of Japan are less than 10^{-4} for the mean core damage annual frequency and less than 10^{-5} for the containment failure annual frequency [7].

To achieve these objectives, the following considerations need to be taken into account in tsunami design:

— The five levels of defence in depth described in SSR-2/1 (Rev. 1) [1] apply to the tsunami protection systems.
— Design basis tsunamis and beyond design basis tsunamis need to be considered.
— Cliff edge effects[1] need to be assessed and prevented at the design basis tsunami and beyond design basis tsunami levels.
— All plant states need to be considered.
— For multiunit sites, each unit needs its own safety systems to address design basis tsunamis and beyond design basis tsunamis.
— A management system and a quality assurance programme of the tsunami design of tsunami protection systems have to be in place.
— Operational effectiveness of non-permanent equipment has to be ensured in challenging circumstances in severe accident conditions.
— Configuration control has to be implemented to support safe operation of the NPP.

2.2. POTENTIAL EFFECTS OF A TSUNAMI ON COASTAL NUCLEAR INSTALLATION SITES

2.2.1. Tsunami hazard phenomena of interest to nuclear power plant sites

According to the literature that was reviewed to arrive at a consensus on the potential effects to be considered, there are several different approaches to categorize and define the effects of a tsunami on a specific nuclear site, depending on the principal effects of interest.

Several different tsunami hazard phenomena are recognized to be significant contributors to the loading conditions to be considered in the design and evaluation of structures, systems and components (SSCs) of NPPs. These

[1] SSR-2/1 (Rev. 1) [1] states:
"A 'cliff edge effect', in a nuclear power plant, is an instance of severely abnormal plant behaviour caused by an abrupt transition from one plant status to another following a small deviation in a plant parameter, and thus a sudden large variation in plant conditions in response to a small variation in an input."

phenomena, and the parameters that express the intensities of these phenomena, are listed below and discussed in more detail in Section 3 (see also Fig. 1):

— Inundation (at the specific location):
  • Tsunami height;
  • Flow depth (inundation depth);
  • Water velocity.
— Runup (at the inundation edge):
  • Runup height;
  • Inundation distance.
— Water level drawdown:
  • Low water level.
— Sediment transport:
  • Bathymetry change;
  • Suspended sediment concentration.

Inundation (Fig. 1(a)) refers to onshore flooding caused by tsunami waves. Key parameters related to inundation are the tsunami height, inundation depth and water velocity, which are defined at specific locations within inundated areas. The tsunami height is the maximum vertical distance from the still water level at tsunami arrival to the elevation of the water surface at the location. Flow depth or inundation depth is the maximum vertical distance upwards from the ground surface or seabed to the elevation of the water surface at the location. Water velocity is the speed of water particles at the location and is typically expressed as a depth averaged value.

Runup (Fig. 1(a)) is the maximum vertical height onshore, above sea level, reached by a tsunami. Its related parameters are the runup height and inundation distance. Runup height is the vertical distance from the still water level at tsunami arrival to the elevation of the edge of the inundation, or the boundary between inundated and non-inundated areas. Inundation distance is the horizontal distance from the shoreline at tsunami arrival to the edge of the inundation.

Water level drawdown (Fig. 1(b)) refers to the lowering of the water surface due to receding of tsunami waves. The related parameter is low water level, which is the maximum vertical distance downwards from the still water level at tsunami arrival to the water surface at the specific location.

Sediment transport (Fig. 1(c)) is the movement of bed material caused by bed friction and turbulence generated by the strong shear of tsunami induced currents. The process of sediment transport is divided into bedload and suspended load. Suspended load is caused by the transport of suspended sediment that is discharged from the seabed to the seawater. Bathymetry change is the elevation

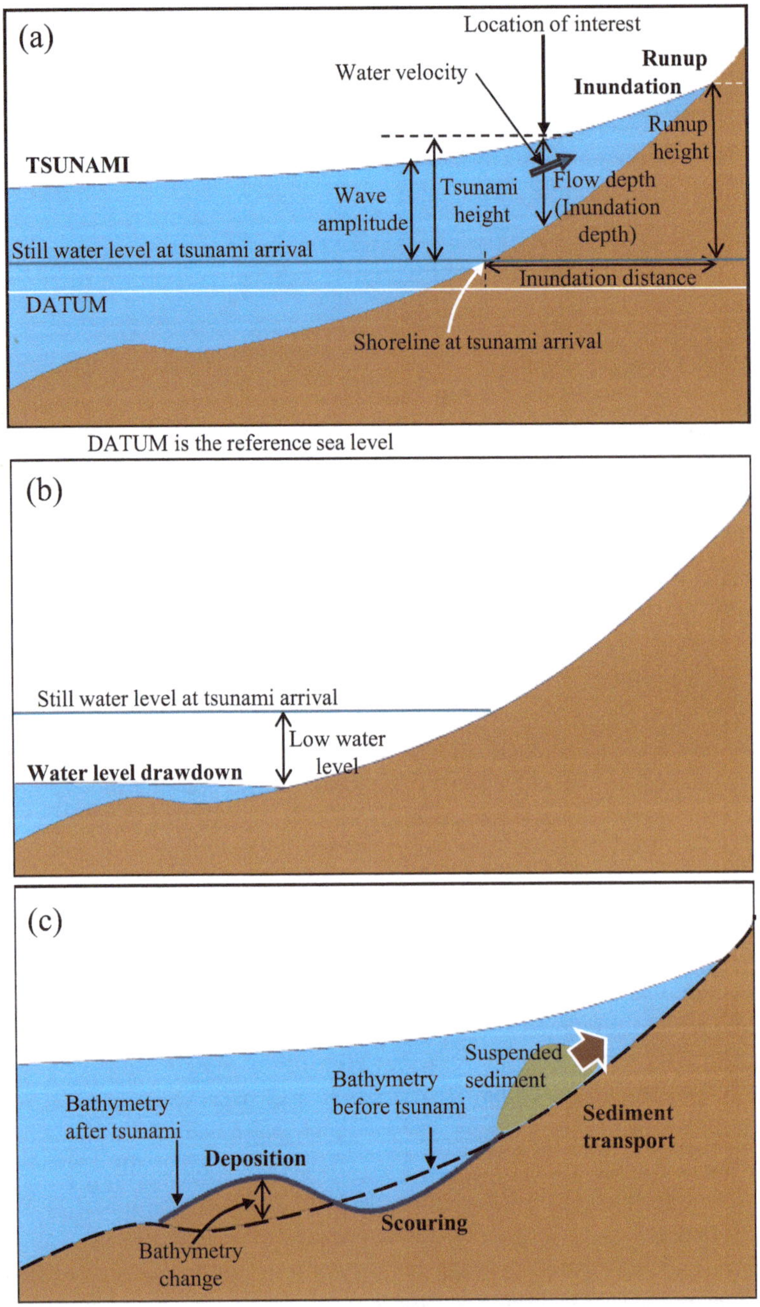

FIG. 1. Tsunami hazard phenomena and related parameters. (a) Inundation and runup; (b) water level drawdown; (c) sediment transport. Figure reproduced from SSG-18 [2].

change of the seabed and a result of sediment transport. Deposition and scouring are the increase and decrease of the elevation of the seabed, respectively.

2.2.2. Summary of loading conditions for nuclear power plant sites

The tsunami hazard phenomena discussed in Section 2.2.1 are recognized to be significant contributors to the loading conditions to be considered in the design and evaluation of SSCs. The important loading conditions include the following:

(a) Hydrostatic loads;
(b) Buoyancy;
(c) Hydrodynamic loads;
(d) Impulsive loads;
(e) Debris impact loads;
(f) Scouring;
(g) Deposition (sediments);
(h) Deposition (debris, other);
(i) Tidal bore;
(j) Drawdown.

In summary, the effects of the tsunami (schematically illustrated in Fig. 2) and its impacts on nuclear installations are as follows:

- Flooding due to tsunami runup and inundation (A), which can have the following impacts on nuclear installations:
  - Loads from wave forces and inundation loads (hydrostatic force, hydrodynamic force, buoyant force) on SSCs (A-1);
  - Immersion of electrical and instrumentation equipment by flooding, leading to loss of function (A-2).
- Dry intakes during drawdown (B), which can lead to:
  - Loss of cooling water.
- Waterborne debris collision (C), which can create:
  - Damage to structures.
- Deposition due to sediment transport and waterborne debris (D), causing:
  - Operating problems to marine utilities;
  - Partial or complete obstruction of intake;
  - Restriction or blocking of flow to the cooling systems, damage to system equipment.
- Scouring due to sediment transport, damage of structure foundations (E).

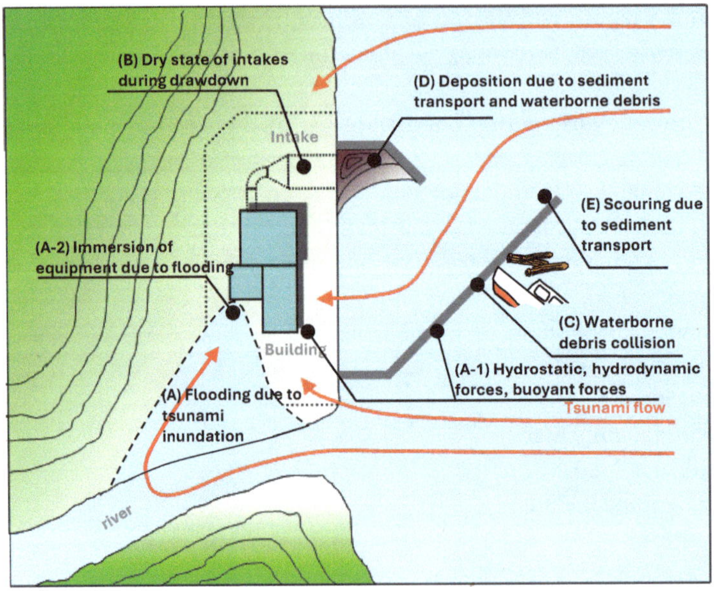

FIG. 2. Effects of tsunamis on nuclear installations (image courtesy of the Nuclear Regulation Authority of Japan, partially revised). The flow directions of the tsunami wave during flooding are indicated by red arrows.

2.3. BASIC APPROACH TO TSUNAMI DESIGN

The focus of this publication is the basic approach to tsunami design for NPPs, but many of the concepts presented here are applicable to other nuclear installations. The elements that have to be considered are as follows:

(a) Hierarchal hazard assessment: Screening of tsunami effects at the site.
(b) Site specific and design specific tsunami assessments:
  — Design basis tsunami;
  — Protection concept;
  — Design principles;
  — Performance criteria.
(c) Tsunami phenomena and tsunami loading conditions:
  — Design basis tsunami;
  — Beyond design basis tsunami.
(d) Tsunami categorization of SSCs.

2.3.1. Hierarchal hazard assessment

The tsunami hazard assessment at an NPP site comprises three stages: (1) screening of tsunami effects; (2) refined assessment of tsunami related phenomena; and (3) implementation of design principles as appropriate. In general, these stages involve modelling of the earthquake sources of concern, initiation and propagation of the wave field from the source to the neighbourhood of the site, and analyses of the various effects of the tsunami phenomena on the site itself.

2.3.1.1. Regional screening

An evaluation needs to be conducted to determine whether the site region is susceptible to tsunamis. For a new site, this evaluation is performed at the site evaluation phase. For existing sites, a re-evaluation of external events will be necessary as part of a periodic safety review or as a result of the occurrence of events, such as the tsunami following the Great East Japan Earthquake. If no tsunami hazard is identified, no further analysis is needed. The finding of no credible tsunami hazard in a given region needs to be supported by extensive region specific evidence, including historical data of all types, evaluations of credible sources of tsunamis worldwide and their lack of consequences to the region, and numerical simulations of worldwide sources verifying no tsunami hazard to the region.

2.3.1.2. Site screening

An evaluation needs to be made to determine whether the NPP is sited and designed in such a way that even if the region is susceptible to tsunamis, the particular site is not adversely affected. Typically, this involves verification of the capacity of the NPP to resist the effects of the tsunami. For example, SSCs located at an elevation above a conservatively established maximum runup elevation for the design basis tsunami are expected to be unaffected. Appropriate margins to the runup levels need to be considered with reference to the beyond design basis tsunami.

2.3.1.3. Site specific assessment and design

Site specific assessment and design are only necessary if the regional and the site screening stages do not screen out the site for further consideration.

2.3.2. Site specific tsunami assessment and nuclear power plant design

The approach to site specific tsunami assessment and design of nuclear installations comprises the following elements and decisions.

2.3.2.1. Definition of the design basis tsunami

The design basis tsunami may be defined deterministically or probabilistically, as a tsunami scenario or using conservative values of the design parameters of interest.

2.3.2.2. Definition of protection concepts

The approach to maintaining the functions of SSCs important to safety is established under the impact of the design basis tsunami. To avoid the impact of the design basis tsunami, the following concepts can be considered.

(a) Dry site concept

SSCs important to safety are constructed above the level of the design basis tsunami loading conditions, with appropriate consideration of the beyond design basis tsunami and possible cliff edge effects. The dry site concept can be applied to the complete installation (i.e. the complete plant site is categorized as being a dry site) or to a portion of the installation (i.e. a subset of all important SSCs, such as the emergency power system) that is located at a high elevation on the site.

For new plants, this can be accomplished, if necessary, by locating the plant at a sufficiently high elevation or by means of construction arrangements that raise the average grade level above the design basis tsunami loading conditions — for example, by constructing an engineered berm on which the NPP SSCs are installed. Such design items are classified as being important to safety and the resulting procedures for design and maintainability over the life of the plant need to be ensured.

For existing plants, any necessary upgrades or retrofits could be placed at elevations satisfying the dry site concept.

(b) Permanent barriers

Levees, sea walls, breakwaters and bulkheads could be constructed external to the plant site or boundary. Systems important to safety, and any components or systems, could be contained in a watertight building or room using watertight doors or other such means. In either case (external barriers or watertight structure), care needs to be taken to implement appropriate design bases for the barriers. In addition to the obvious tsunami loading conditions, seismic design needs to be taken into account, especially for sites susceptible to tsunamis during and after earthquakes, but also, in general, to ensure that the barriers' capability is not degraded by other external events. These barriers are classified appropriately, designed for appropriate load combinations and maintained over the plant life. Maintenance requirements include periodic inspections, monitoring and maintenance of the external barriers — that is, both the barriers that are under the responsibility of the plant operating organization and those that are not. The permanent external barriers that are under the responsibility of the plant operating organization are considered items important to safety.

In the same manner as in the dry site concept, permanent barriers can be implemented either for the entire site (e.g. a sea wall blocking all paths of water intrusion) or for a portion of the installation (e.g. a subset of all important SSCs).

(c) Design of SSCs

It is also possible to install SSCs important to safety that can maintain functionality even when subjected to the impact of the tsunami, such as watertight motors and pumps.

(d) Combinations of protection concepts

Combinations of the above three protection concepts are acceptable, especially for existing plants, where it may not be feasible to develop a totally dry site concept, but upgrades or retrofits may address a combination of the protection concepts. For example, if watertight doors for a building are not adequate to resist the hydrodynamic loads associated with the tsunami, a sea wall may be constructed specifically to reduce — but not eliminate — the hydrodynamic loads on the watertight doors. Thereby, the combination of two concepts protects the items inside the building from tsunami induced damage.

(e) Defence in depth

Defence in depth could be addressed by implementing a multiple protection concept. For example, one level of defence is a permanent barrier, such as a sea wall. Another level of defence is watertight doors to the diesel generator building; such doors are another permanent embedded barrier, but it is necessary to ensure that they will remain closed during a tsunami. An additional level of defence is a new redundant emergency power train constructed at a high elevation, either on a natural hill or a constructed engineered berm.

(f) Standard design basis tsunami and beyond design basis tsunami

For a standard NPP design to be placed in areas and on sites that are susceptible to tsunami hazards, a standard design basis tsunami and beyond design basis tsunami could be defined for design considerations. The standard design could adopt a hierarchy of protection measures to be applied in the design process or offer optional features to protect the plant against the design basis tsunami and the beyond design basis tsunami. The licensee defines the standard design basis tsunami and the beyond design basis tsunami, which are treated as design conditions and beyond design conditions. With reference to both design bases, the approach to design could be as follows:

— The complete plant is located at a dry site (i.e. at an elevation above the design basis tsunami loading conditions), with adequate margin to meet beyond design basis tsunami conditions with confidence.
— A combination of dry site concepts with permanent barriers is developed.
— A combination of dry site, permanent barriers and functional assurance of equipment in a wet environment is developed as the standard design.

In all options, design robustness against tsunamis needs to be ensured.

2.3.2.3. Tsunami design principles and defence in depth principles

Assuming that tsunami modelling of the site has been performed and the design basis tsunami and the beyond design basis tsunami have been established, the following processes are carried out:

(a) Dry site concept for the design basis tsunami and the beyond design basis tsunami. The complete nuclear installation is evaluated to check whether a significant safety margin exists. Considerations include the layout of the plant and safety related SSCs. If a significant margin exists to the beyond design basis tsunami, no additional measures may be necessary.
(b) Permanent barriers. The following considerations apply:
    (i) Permanent barriers are designed for all the effects of the phenomena of the design basis tsunami — inundation (tsunami wave height, flow depth and water velocity), runup, low water level drawdown and sediment transport (see loading conditions in (iii)).

(ii) Permanent barriers protecting SSCs important to safety are evaluated against the beyond design basis tsunami conditions.

(iii) Loading conditions to be considered in the design of permanent barriers are categorized according to the potential failure modes leading to cliff edge effects. For example, a sea wall for hydrostatic loads, buoyancy, hydrodynamic loads, impulsive loads, debris impact loads and drawdown is designed. Current design codes for these loading conditions require the sea wall to be designed so as to behave in a ductile manner, which in turn provides assurance that there will be significant capacity to resist beyond design basis tsunami loads without failure. Other loading conditions imposed on the sea wall, such as over-topping, scouring or other sources of soil failure (e.g. leading to liquefaction), may lead to cliff edge effects. In such cases, loading conditions are identified for the evaluations described in (v).

(iv) Defence in depth may be needed for permanent barriers. A prudent design concept is to have a minimum of two independent redundant permanent barriers (e.g. sea wall and watertight buildings and enclosures for systems, components, equipment and distribution systems important to safety).

(v) Progressive system collapse evaluations are performed. Sequential loss of permanent barriers is assumed, and the consequences are assessed from the standpoint of overall NPP safety. The tsunami protection measures are considered as a system.

(vi) The permanent external barriers are considered as items important to safety.

(vii) SSCs important to safety and designed to operate under tsunami loading conditions are labelled as functional barriers. For example, passive mechanical components designed to maintain structural integrity when subjected to tsunami loading conditions (e.g. piping, tanks); active components designed to operate under tsunami loading conditions (e.g. submergible pumps); electrical equipment with water resistant capability to permit functionality; infrastructure (e.g. cables, connections) qualified to operate when submerged or subjected to a water environment; intake facilities, including seawater pumps designed to maintain structural integrity and operability when subjected to tsunami loading conditions, in particular drawdown that will cause inlets to be dry for a period of time.

2.3.2.4. Establishment of design and performance criteria for nuclear installations

The sole purpose of the tsunami specific designed elements is to protect the SSCs important to safety of the nuclear installation from damage or loss of function due to the tsunami loading conditions. A tsunami sea wall is an example. These elements are typically categorized as permanent barriers.

The tsunami design is applicable to the following SSCs of the nuclear installation: (a) SSCs classified as important to safety and necessary to ensure and maintain the safety of the nuclear installation; and (b) SSCs that provide tsunami protection for the SSCs in (a). Examples are permanent barriers, such as watertight penetrations or doors in walls of structures housing systems and components important to safety; and systems and components qualified to operate in the environment created by tsunami loading conditions, such as submersible pumps operating when inundated by water.

Categorization of the items important to safety for tsunamis and the tsunami protection system are discussed in Section 2.3.4. Briefly, elements of the tsunami protection system and a subset of SSCs important to the safety of nuclear installations are categorized as Tsunami Category 1 (TC-1). These items are designed and/or evaluated specifically for the design basis tsunami and the beyond design basis tsunami. The definition of TC-1 SSCs starts with all SSCs of the nuclear installation categorized as seismic category 1, which then are reduced to a subset supplemented by permanent barriers and functional barriers that are specific to the tsunami design.

Tsunami Category 2 (TC-2) items are those that are not part of the tsunami protection system but may have an effect on the safety of the nuclear installation or on the tsunami protection system.

2.3.2.5. Other design considerations

Local and distant sources of tsunamis impose different initial conditions on NPPs, which should be taken into account.

Local tsunamis are produced by earthquakes occurring close to the site in many cases. Consequently, it is expected that earthquake ground motion will be experienced at the site. Depending on the earthquake source parameters that are likely to produce the tsunami (e.g. magnitude, distance, fault rupture characteristics), the effects of shaking on the NPP site may be: (a) ground motion exceeding the design basis earthquake (DBE) ground motion, causing no damage, minimal damage or significant damage to SSCs; (b) ground motion of amplitude lower than the DBE and greater than the operating basis earthquake, causing no or minimal damage to SSCs important to safety, but leading to automatic scram or manual shutdown of the operating NPP; (c) ground motion of amplitude lower than the operating basis earthquake, but significant enough to cause automatic scram or leading to manual shutdown.

In all these cases, a pre-existing condition exists, which is taken into account in the design and evaluation phase. This affects assumptions associated with the operational states of NPP load combinations.

Distant tsunamis differ from local tsunamis in that earthquake vibrations are not experienced at the site. In the case of a distant tsunami, if combination with a local tsunami is considered, the NPP may be in a state of hot or cold shutdown due to automatic or manual shutdown resulting from vibratory ground motion experienced at the site or from advance tsunami warning.

2.3.3. Tsunami loading conditions

2.3.3.1. Design basis tsunami

A design basis tsunami can be defined by a set of tsunami phenomena, described in Section 2.2.1, or by a set of tsunami loading conditions, described in Section 2.2.2.

References [2, 3, 8] provide background, methodologies and approaches for defining the tsunami hazard at an NPP site. The following are basic concepts to be considered:

(a) Definition of control point(s). The control point may be one location or a series of locations offshore and represents the transition point between modelling the propagation of the tsunami from the source to a location where non-linear onshore modelling is initiated. The control point is located where: (a) the linear long wave theory applies; and (b) reflected waves from the coast are not significant.
(b) Probabilistic definition of the tsunami hazard [2]. Important aspects of the overall process include the following:
   (i) Simulations of individual scenario tsunamis are performed in the probabilistic analyses. Each scenario is associated with a source region and with a frequency of occurrence for a probabilistic tsunami hazard assessment.
   (ii) At a specified control point, analysis of the scenario tsunami is transferred from the overall propagation analysis to the local site specific analysis, taking into account the site specific characteristics.
   (iii) For a full probabilistic analysis, variability (aleatory and epistemic uncertainty) in all the models, as well as in the parameters associated

with both linear and non-linear analyses, is explicitly considered. The end results are the probability distributions of the tsunami phenomena described in section 2.2.1 and the tsunami loading conditions described in Section 2.2.2. A probability distribution could be conditional on a tsunami occurring or unconditional due to convolving over the frequency of occurrence of all tsunamis that produce the tsunami phenomenon, the tsunami loading condition or both.

(iv) An important consideration of these results is that individual tsunami phenomena may be correlated and individual tsunami loading conditions may be correlated. For example, a large tsunami height may or may not be correlated with a large wave velocity. Similarly, a large hydrodynamic load may or may not be correlated with a large hydrostatic load. Each scenario may have unique correlative properties amongst the resulting parameters. For hydrodynamic loads, wave height and flow velocity are correlated, and the parameter of interest is their combination (i.e. momentum flux; see Section 5).

(c) Performance of a series of simulations of individual tsunami scenarios. These scenarios are selected probabilistically or deterministically, as follows:

(i) Probabilistically defined scenarios that are not conditioned by their frequencies of exceedance result in unconditional distributions of the tsunami phenomena and tsunami loading conditions. From these probability distributions, values of the tsunami phenomena and tsunami loading conditions can be associated with a frequency of exceedance (e.g. 10^{-4}, 10^{-5} and 10^{-6} per annum).

(ii) Deterministically selected scenarios may be based on limited historical data, sensitivity studies, numerical simulations, expert opinion and other considerations. The end product of the simulations is a tabulation of individual tsunami loading conditions ranked from high to low, but without regard for correlation between scenarios. A set of loading conditions are produced that are eventually associated with each of the TC-1 SSCs.

(d) For the design basis tsunami, selection of loading condition values for the design of TC-1 SSCs and the evaluation of TC-2 SSCs, according to the following criteria:

(i) If the number of tsunami simulations is large enough, each of the tabulations mentioned in (c)(ii) could be interpreted probabilistically and the designer could select a probability of exceedance value to define the design basis tsunami quantities. For example, a value of the median plus one standard deviation could be selected for each of the tsunami loading conditions, conditional on a tsunami occurrence.

Alternatively, the designer could target a probability of exceedance value such as 10^{-4} or 10^{-5} per annum.

(ii) Regardless of the number of simulations, the designer may decide to select the maximum value of the tabulation for the design basis tsunami quantities.

(iii) A safety factor may be added to some or all the tsunami loading conditions, depending on the loading condition, the SSC to which it applies, the nature of the consequences of exceeding the design basis (e.g. a cliff edge effect) and other considerations.

The design basis tsunami loading conditions are specified for all TC-1 SSCs by taking into account the location and topography of the NPP site.

2.3.3.2. Beyond design basis tsunami

Paragraph 5.21A of SSR-2/1 (Rev. 1) [1] states:

"The design of the plant shall also provide for an adequate margin to protect items ultimately necessary to prevent an early radioactive release or a large radioactive release in the event of levels of natural hazards exceeding those considered for design, derived from the hazard evaluation for the site."

A beyond design basis tsunami is defined as a set of tsunami phenomena or tsunami loading conditions that are greater than those of the design basis tsunami and that are suitable for assessing the safety margin beyond the design basis tsunami conditions.

Paragraph 3.24 of SSG-68 [3] states that: "The definition of the beyond design basis external event loading conditions is inherently connected with the performance and acceptance criteria for SSCs and the nuclear installation."

In principle, the beyond design basis tsunami should challenge the NPP, especially tsunami phenomena and tsunami loading conditions that could lead to cliff edge effects. Logically, the criteria for defining the beyond design basis tsunami could be based on the tsunami probabilistic risk assessment plant metrics and possibly include other measures, such as a safety margin.

Paragraph 3.24 of SSG-68 [3] further states that: "methodologies to evaluate beyond design basis external events may be performed by means of a best estimate approach (which is relaxed compared with design methods and acceptance criteria relating to material properties)."

To assess the margins and evaluate cliff edge effects, one of the following methods for defining the beyond design basis external event (BDBEE) loading conditions is used:

(a) Defining the BDBEE loading conditions by applying a factor to its loading conditions [3];
(b) Defining the BDBEE loading conditions on the basis of the probabilistic hazard evaluation;
(c) Defining the BDBEE loading conditions as the maximum credible hazard severity.

2.3.3.3. Design extension conditions

SSR-2/1 (Rev. 1) [1] defines DECs as:

> "Postulated accident conditions that are not considered for design basis accidents, but that are considered in the design process of the facility in accordance with best estimate methodology, and for which releases of radioactive material are kept within acceptable limits."

DECs encompass scenarios where significant fuel degradation does not occur and scenarios involving core melting. The principal technical concept in considering DECs is to ensure that the design of the NPP either prevents accident conditions not classified as design basis accident conditions or mitigates their consequences to an extent that is reasonably practicable. SSR-2/1 (Rev. 1) [1] states:

> "The design of the plant shall also provide for an adequate margin to protect items ultimately necessary to prevent an early radioactive release or a large radioactive release in the event of levels of natural hazards exceeding those considered for design, derived from the hazard evaluation for the site."

Reference [8] discusses DECs and BDBEEs and explains the usefulness of familiarization with the concepts of BDBEE (specifically, beyond design basis tsunami) and DECs as part of the design and assessment of an NPP (or other nuclear installations).

2.3.3.4. Severe accident prevention and management

Measures to prevent and mitigate severe accidents have been implemented at NPPs in many Member States following the March 2011 accident at the Fukushima Daiichi NPP. Requirements 58, 68 and 80 and the associated paragraphs in SSR-2/1 (Rev. 1) [1] acknowledge the usefulness of non-permanent equipment being added as complementary to the fourth level of defence in depth.

IAEA Safety Standards Series No. SSG-67, Seismic Design for Nuclear Installations [9], and Ref. [10] detail an approach used broadly to supplement existing safety systems in NPPs.

SSR-2/1 (Rev. 1) [1] specifies non-permanent equipment, including equipment for restoring the capability to remove heat from containment, restoring the necessary electrical power supply and ensuring sufficient water inventory for the long term cooling and shielding of spent fuel. However, the scope of non-permanent equipment is not limited to these examples. In addition, SSR-2/1 (Rev. 1) [1] describes non-permanent equipment as follows:

(a) "The design shall also include features to enable the safe use of non-permanent equipment for restoring the capability to remove heat from the containment" (para. 6.28B).

(b) "The design shall also include features to enable the safe use of non-permanent equipment to restore the necessary electrical power supply" (para. 6.45A).

(c) "The design shall also include features to enable the safe use of non-permanent equipment to ensure sufficient water inventory for the long term cooling of spent fuel and for providing shielding against radiation" (para. 6.68).

Non-permanent equipment, with its flexibility to respond to various DEC scenarios, can be an effective measure in accident management to control the release of radioactive material to the environment. References [8, 11] give detailed examples of non-permanent equipment and its use for accident management.

2.3.4. Tsunami categorization for structures, systems and components

Categorization of SSCs for tsunami design, tsunami protection and tsunami evaluation is an essential part of the tsunami design process. One approach is to identify TC-1 and TC-2 SSCs.

2.3.4.1. Tsunami Category 1 structures, systems and components

Theoretically, all SSCs important to safety[2] in a nuclear installation are categorized as TC-1. This is very similar to the treatment of seismic events. Although there is a parallel in the tsunami categorization of SSCs with their seismic categorization, the former differs significantly in that a tsunami event as a common cause external event does not directly or indirectly affect all SSCs important to safety simultaneously, as seismic events do. The categorization procedure is as follows:

(a) The first step in the categorization process is to identify buildings and structures that will serve as tsunami protection for SSCs important to safety located within. The protection concept is to apply the dry site, the permanent barrier and/or the functional barrier concepts to prevent failure of the building or a portion thereof, such that there is no water intrusion into the building envelope. The building is categorized as TC-1. In general, the SSCs important to safety located within the building are assumed to be protected and are not identified as TC-1. The following examples apply:
   (i) An example of tsunami protection is the typical containment building or reactor building. If the principle of defence in depth is applied, a sea wall protecting the site can be envisioned as the first level of protection, with watertight doors, penetrations and hatches installed in the building as the second level of protection. Then, all SSCs important to safety located in the building are not TC-1. This step eliminates hundreds or even thousands of SSCs from further evaluation.
   (ii) A second example is a turbine building where SSCs important to safety are not normally installed, but that may nevertheless house them. In such a case, the first level of protection is the sea wall as before. However, the second level of protection may not serve as protection of the complete turbine building envelope, which may be too difficult to ensure owing to all the accesses and penetrations. In this case, the second level of protection may be internal to the turbine building, such as ensuring that rooms or compartments housing SSCs important to safety are isolatable and watertight (permanent barrier), or ensuring that the SSCs can perform their functions in an inundated condition.

---

[2] The nomenclature of SSR-2/1 (Rev. 1) [1] is adopted herein with respect to the safety classification of SSCs. This standard categorizes plant equipment into items important to safety and items not important to safety [1]. It further categorizes items important to safety into safety related items and safety systems (in this context, an item is a structure, system or component) [1].

Step 1 is the first level screen (screen out) of SSCs important to safety. The end product of Step 1 is a list of buildings and structures that are categorized as TC-1 and need to be designed for the design basis tsunami and evaluated for the beyond design basis tsunami.

(b) The second step is to identify SSCs important to safety located outside of the buildings; that is, in the yard of the site. These SSCs need to be protected by implementing one or more protection concepts. In this case, the barriers and/or the SSCs important to safety are identified as TC-1. Step 2 is the second level screen (screen in) of SSCs important to safety located outside of buildings. The end product of Step 2 is a list of SSCs important to safety located outside of buildings that are TC-1. These SSCs need to be designed for the design basis tsunami and evaluated for the beyond design basis tsunami, or protected by the tsunami protection system.

2.3.4.2. Tsunami Category 2 structures, systems and components

TC-2 items are those items that are not part of the tsunami protection system but may have an effect on the nuclear installation or on the tsunami protection system. Examples include items that could become flotsam and consequently become waterborne missiles impacting SSCs important to safety, and sources of fire, such as flammable fluid contained in tanks that could fail and subsequently ignite, causing fire or explosive damage.

2.3.4.3. Subcategorization of Tsunami Category 1 and 2 items

TC-1 items that are part of the tsunami protection system can be subcategorized on the basis of their purpose and function. One such designation is the following:

(a) External permanent barriers, whose sole purpose is as elements of the tsunami protection system (e.g. a tsunami sea wall). These items could be designated TC-1(e).
(b) Incorporated permanent barriers, whose purpose may be a combination of operational requirements and a part of the tsunami protection system (e.g. a watertight door on a TC-1 building). These items could be designated TC-1(i).
(c) Items designed or qualified to operate in wet conditions (e.g. submergible pump). These items could be designated TC-1(t).

Similar subcategorizations could be implemented for TC-2 items, if appropriate.

3. TSUNAMI PHENOMENA AND LOADING CONDITIONS

3.1. TSUNAMI PHENOMENA

Section 2.2.1 introduced the tsunami phenomena of interest to the definition of a design basis tsunami and beyond design basis tsunami. These phenomena are discussed in more detail in this section.

The following tsunami hazard phenomena are recognized to be significant contributors to defining the loading conditions to be considered in the design and evaluation of NPP SSCs:

(a) Tsunami height. The height of water is defined at the locations of interest relative to the elevation of water or the ground elevation at this location. For example, it can be defined at the following locations:
   (i) At the control point (Section 2.3) offshore. The tsunami height at the control point is one of the tsunami parameters at the hand-off between modelling the propagation of the tsunami from the seismic source to a location where non-linear onshore modelling is initiated.
   (ii) At the location where a permanent barrier is to be constructed or installed, such as a sea wall or a watertight door in a building. The tsunami height at such a location is a design parameter for a design basis tsunami and an evaluation parameter for a beyond design basis tsunami.

(b) Flow (or inundation) depth. Flow depth is the water depth at the location of interest. As well as the tsunami height, it can be used to determine the inundation and intrusion of SSCs installed at the location.

(c) Water velocity. Water (or flow) velocity is defined at locations of interest. Water velocity parallels the definition of tsunami height; it comprises a correlated pair of tsunami parameters used in site specific tsunami assessments and in the definition of tsunami loading conditions. Momentum flux (see Section 5) defines the hydrodynamic force applied to SSCs.

(d) Runup height. Runup height is the height above the shoreline at the runup location (runup elevation minus still water level at tsunami arrival). This serves as a criterion for achieving the dry site concept, in conjunction with considerations for conservatism.

(e) Inundation distance. Inundation distance is the horizontal distance from the shoreline corresponding to the still water level at tsunami arrival to the location of interest.

(f) Low water level. Low water level due to receding tsunami waves (or drawdown) causes additional hydrodynamic loads on SSCs, loss of cooling water and dry intake causing failure of pumps, such as seawater pumps.
(g) Sediment transport. Deposition caused by sediment transport may obstruct the water intake, and scouring due to sediment transport may impact offshore breakwaters.

3.2. DIRECT AND INDIRECT EFFECTS OF TSUNAMIS ON THE SITE

Phenomena of importance to the tsunami design of the NPP can have direct effects and indirect effects.

(a) Direct effects include inundation (flooding due to tsunami runup), hydrostatic forces, wave force (hydrodynamic direct impact and suction forces) and buoyancy; these are sometimes referred to as 'tsunami impacts'.
(b) Indirect effects include flotsam collisions, sand deposition, scouring, morphology, fires and other events that may be triggered by a tsunami; these are sometimes referred to as 'tsunami associated events'.

3.3. TSUNAMI LOADING CONDITIONS FOR THE NUCLEAR POWER PLANT SITE

3.3.1. Hydrostatic loads

Hydrostatic forces occur when standing water or slowly moving water exerts pressure on a structure (e.g. building), a structural element (e.g. building wall or roof, sea wall, penetration), a geotechnical structure or element (e.g. berm, channel), and mechanical and electrical components (e.g. tanks). Hydrostatic forces act perpendicularly to the surface of the item. Net hydrostatic forces result from an imbalance of pressure due to different water levels on opposite sides of the item of interest. Vertical hydrostatic forces contribute to offsetting buoyancy forces.

Hydrostatic forces are calculated by various formulas, taking into account the density of the water, including sediments.

3.3.2. Buoyancy

Buoyancy forces are vertical, upward acting hydrostatic forces. The total buoyancy force acts at the centroid of the volume displaced by a structure, structure element, geotechnical element, mechanical component, etc. The total buoyancy force equals the weight of the displaced water.

Buoyancy forces are a concern for structures that are part of the tsunami protection system and items that could become missiles carried by tsunami waves and impacting tsunami protection system components and NPP SSCs.

3.3.3. Hydrodynamic loads

Hydrodynamic forces are caused by the water flow impacting directly the structure, structural element, geotechnical structure or element, and mechanical/ electrical components. They are a function of fluid density (including suspended particles), flow velocity and impacted item geometry. They are caused by steady state and impulsive type loading conditions, and include suction forces generated during receding of the water.

3.3.4. Impulsive loads

Impulsive forces are caused by the initial impact of the tsunami waves. They are rapidly applied forces and need to be treated as such in the design and evaluation phases; that is, by applying appropriate dynamic amplification factors and acceptance criteria.

3.3.4.1. Debris impact loads

Waterborne debris that is transported to the NPP site imposes dynamic loading conditions on structures, structural elements, geotechnical structures or elements, and mechanical and/or electrical components. Examples of waterborne debris include boats, shipping containers, buildings, land vehicles (e.g. automobiles, trucks, trailers), lumber and trees. Assumptions are needed to define the impact loading conditions of debris. For design and evaluation purposes, waterborne debris can be addressed using an approach similar to that used for tornado or hurricane borne missiles. Waterborne debris may be transported from outside the plant boundary to the NPP site or may be generated within the site.

3.3.4.2. Scouring

Scouring is a localized loss of soil in the neighbourhood of a foundation supporting a structure (e.g. building), structural element (e.g. sea wall), geotechnical structure or element (e.g. berm, channel, breakwater) and mechanical and/or electrical components (e.g. tanks). Two primary scouring mechanisms occur during a tsunami event: (a) shear induced scouring, which consists of soil transport due to the flow velocity and is similar to scouring caused by a storm surge or other flooding phenomena; and (b) liquefaction induced scouring, which results from rapid drawdown as water recedes. Scouring can lead to significant soil failure modes and needs to be evaluated by senior, experienced geotechnical engineers.

3.3.4.3. Deposition (sediments)

Suspended particles in the water may be deposited on structures, structural elements, geotechnical structures or elements, and mechanical and/or electrical components, creating added loading conditions and operability issues, such as blocking of intakes, screens, fouling of active mechanical components and malfunction of electrical equipment, including cables and/or cable connections.

3.3.4.4. Deposition (debris, other)

Waterborne debris may be deposited on, in front of, and behind structures, structural elements, geotechnical structures or elements, and mechanical and/or electrical components, limiting their ability to perform their required functions. On-site or off-site items may be the source of such debris.

3.3.4.5. Tidal bore

See Sections 3.3.3 and 3.3.4.

3.3.4.6. Drawdown

Receding water leads to added direct loading conditions due to hydrodynamic effects, scouring, soil failures (e.g. liquefaction) and functional failure of mechanical components (e.g. pumps running dry).

3.3.4.7. *Inundation*

Inundation is a tsunami phenomenon that has additional special significance. It is the source of many tsunami loading conditions. It is also the potential source of water ingress into TC-1 building housing systems, components, equipment and distribution systems important to safety. A part of the design process of the tsunami protection system is to perform an assessment of the inundation effects, as follows:

(a) Evaluate inundation depths on the basis of the tsunami inundation distribution.
(b) Assess all sources of potential water penetration into buildings; for example, through doors, penetrations and vents.

Five different tsunami hazard phenomena (see Section 2.2.1) are recognized to be significant contributors to defining the loading conditions to be considered in the design and evaluation of SSCs. Table 1 provides a correlation of these phenomena with loading conditions. Table 2 lists the potential effects of tsunami phenomena on SSCs.

TABLE 1. IMPORTANT TSUNAMI HAZARD PHENOMENA IN VARIOUS LOADING CONDITIONS
(*adapted from Ref.* [12])

Loading conditions	Important tsunami hazard phenomena					
	Inundation			Runup	Water level drawdown	Sediment transport
	Tsunami height	Flow depth	Water velocity	Runup height	Low water level	
Hydrostatic loads	X	X	X	X		
Buoyancy	X	X				
Hydrodynamic loads	X		X			
Impulsive loads			X			
Debris impact loads			X			
Scouring			X			X
Deposition (sediments)			X			X
Deposition (debris, other)			X			X
Tidal bore	X	X	X	X		
Drawdown					X	

TABLE 2. TYPICAL EFFECTS OF TSUNAMI PHENOMENA ON STRUCTURES, SYSTEMS AND COMPONENTS

Facility			Damage mode	Related tsunami parameter
Location	Type	SSCs typically affected		
Outdoor	Civil engineering structure	Breakwater Intake/discharge facility Foundation	Structural damage by wave forces and inundation loads	Flow velocity and inundation depth
			Structural damage by debris collision	Flow velocity
			Functional damage by sediment and debris deposition (intake facility)	Flow velocity
			Structural damage by scouring	Flow velocity and inundation depth
	Construction engineering structure	Building	Structural damage by wave forces and inundation loads	Flow velocity and inundation depth
			Structural damage by debris collision	Flow velocity
	Mechanical equipment	Tank Pump Valve Piping	Structural damage by wave forces and inundation loads	Flow velocity and inundation depth
			Structural damage by debris collision	Flow velocity
			Functional damage by drawdown (seawater pump) or sand intrusion	Water level at a pump (time history)
	Electrical and instrumentation equipment	Transformer Pump motor Valve actuator	Functional damage by immersion	Inundation depth

TABLE 2. TYPICAL EFFECTS OF TSUNAMI PHENOMENA ON STRUCTURES, SYSTEMS AND COMPONENTS (cont.)

Facility			Damage mode	Related tsunami parameter
Location	Type	SSCs typically affected		
Indoor	Mechanical equipment	— Tank — Pump — Valve — Piping	Structural damage by inundation (inside building)	Inundation depth at leaking area into building (time history) Inundation height
	Electrical and instrumentation equipment	— Pump motor — Valve actuator	Functional damage by immersion	Inundation depth at leaking area into the building (time history) Inundation height

4. TSUNAMI DESIGN PROCESS

The tsunami design process is applicable to new NPPs and to upgrades or retrofits for existing plants, with some adjustments.

Section 2.3.2 presents the overall approach to defining the design basis tsunami. As stated, the design basis tsunami may be defined in terms of tsunami phenomena or tsunami loading conditions or a combination of both.

4.1. PHASE 1

In phase 1, the design basis tsunami loading conditions for TC-1 and TC-2 SSCs are determined using site specific analyses. The starting point for the tsunami simulations is at the control point, as described in Section 3.3. A series of simulations are performed. Each simulation is initiated at the control point; the site topography (natural and human made elements) is modelled; and the tsunami wave is propagated onto the site, generating values of the tsunami phenomena and/or tsunami loading conditions. As stated in Section 3.3, the design values for tsunami phenomena and/or tsunami loading conditions are selected on the basis of the values derived from the simulations.

The end product of the phase 1 tsunami analyses is tsunami phenomena design parameters, such as runup, extent and height of inundation, and low water level. These values lead to preliminary design concepts for the tsunami protection system. Simplified assumptions may be used in lieu of performing simulations.

4.2. PHASE 2

In phase 2, the preliminary design of the tsunami protection system is prepared, adhering to the design philosophies of dry site, permanent barriers, functional barriers and a combination of these three. The preliminary design is focused on the site and layout of the NPP and the tsunami protection system. In generating the preliminary design, relevant specific tsunami design principles for SSCs important to safety as specified in Section 2.3.2 (e.g. defence in depth principles for tsunami protection systems) are adhered to.

The end product of the preliminary design comprises the layout and implementation of the concepts of dry site, permanent barriers, functional barriers and combinations thereof to address tsunami phenomena, including defence in depth principles.

4.3. PHASE 3

In phase 3, the tsunami analyses of phase 1 are repeated to determine whether the conceptual design incorporating dry site, permanent barriers, functional barriers and combinations thereof satisfies the design performance criteria. If it does, the next phase can begin. If not, the preliminary design is adjusted, and phases 1 and 2 are repeated until the results using the preliminary design concepts satisfy the design performance criteria. This iteration is necessary only if permanent barriers are introduced that significantly change the onshore flow characteristics of the tsunami. The optimization of different concepts is encouraged.

The end products of phase 3 are: the final preliminary design of the tsunami protection system, comprising the layout and implementation of the concepts of dry site, permanent barriers and functional barriers to address tsunami phenomena; a tsunami protection system that adheres to the defence in depth principle; and elements of the tsunami protection system that are identified.

4.4. PHASE 4

In phase 4, the final design of the tsunami protection system is initiated, taking into account the TC-1 and TC-2 SSCs.

The dry site concept is taken into consideration. For new plants and for upgrades or retrofitting of existing plants, the finished grade level of the plant, or a portion thereof, may be placed at a level conservatively above the design basis tsunami runup level, as determined in phases 1–3. In this case, the remaining design and evaluation tasks are: to assess whether there is any conservatism in the dry site configuration; to assess the defence in depth if the design basis tsunami phenomena are exceeded; and to evaluate the beyond design basis tsunami loading conditions.

The defence in depth approach to determining the design loading conditions for all TC-1 SSCs is implemented. This approach entails progressive tsunami protection system collapse evaluations. The sequential loss of permanent barriers is assumed and the consequences are assessed from the standpoint of SSCs important to safety and of overall NPP safety.

For SSCs important to safety, the proposed tsunami defence in depth approach to protection is defined. A tabulation of tsunami protection system elements for SSCs important to safety by individual SSC, groups of SSCs (such as all SSCs important to safety in the reactor building) or other grouping is helpful.

For the first line of defence, the tsunami loading conditions for the geotechnical, structural, mechanical and electrical design are defined. For illustration purposes, the first line of defence is assumed to be a civil structure, such as a sea wall, that is subjected to the direct impact of the tsunami waves, which produce direct impacts such as hydrostatic loads, buoyancy, hydrodynamic loads, impulsive loads, debris impact loads and drawdown loads. Indirect impact due to other loading conditions, such as scouring, is also considered.

Given these loading conditions, a preliminary design is prepared, taking into account all tsunami loading conditions (see Sections 4 and 5). Pre-existing conditions, specifically seismic loads due to the near-field earthquake assumed to produce the tsunami, are taken into account. In addition, other loading conditions assumed to act simultaneously with the tsunami loads are included in load combinations, for example, aftershocks from the near-field earthquake that is assumed to be the origin of the tsunami.

Assuming that the first line of defence fails, the tsunami propagation analysis leading to calculated tsunami phenomena is repeated using the resulting tsunami loading conditions for each of the second level defence in depth elements of the tsunami protection system, of which there could be many (e.g. additional water diverting walls on site, watertight doors/penetrations). The tsunami loading conditions for all tsunami protection system elements are generated. The second level defence in depth tsunami protection system elements (e.g. geotechnical, structural, mechanical, electrical) are designed.

The design process is continued for all tsunami protection system elements. Evaluations are conducted of the performance of the NPP when subjected to beyond design basis tsunamis, as defined to confirm the effectiveness of the designs for the design basis tsunami. Appropriate quality assurance of the analysis and design, quality control of the construction, and maintenance are performed during the operational lifetime of the NPP.

5. DESIGN OF EXTERNAL AND INCORPORATED BARRIERS

The functional performance criteria for geotechnically, structurally or mechanically designed permanent barriers are that the barriers maintain their design function when subjected to the design basis tsunami loading conditions.

The geotechnical aspects of structural design (i.e. the foundation design of structures, such as buildings and sea walls) are to provide foundation support without degradation due to tsunami loading conditions. The design functions for structures as a function of their purpose are as follows:

(a) Tsunami specific designed elements, whose sole purpose is to protect the SSCs important to safety of the nuclear installation from losing their functions owing to the tsunami loading conditions. These are permanent barriers categorized as TC-1(e); a tsunami sea wall is an example. These TC-1(e) permanent barriers are designed to be within the code of allowable stresses for the applicable tsunami loading conditions. The barriers are evaluated for beyond design basis tsunamis, and the evaluation results are used to confirm maintenance of the safety functions of the SSCs important to safety.
(b) Permanent barriers, such as watertight penetrations or doors, installed in the exterior or interior of a building housing systems and components important to safety, are categorized as TC-1(i). These TC-1(i) permanent barriers are designed to be within the code of allowable stresses for the same tsunami loading conditions as for TC-1(e) permanent barriers. However, in the case of TC-1(i) permanent barriers installed in a building, tsunami loading conditions such as hydrodynamic loads, impulsive loads, debris impact loads and scouring may potentially be mitigated, if these permanent barriers are protected by TC-1(e) and/or other TC-1(i) permanent barriers. The barriers are evaluated for the beyond design basis tsunami condition, and the evaluation results are used to confirm maintaining the safety functions of the SSCs important to safety.

The applicability of the information provided in this section extends to buildings and structures that are subject to direct tsunami external force (mainly TC-1(e) and some TC-1(i)), to buildings and structures protecting necessary facilities during a tsunami (TC-1(i)) and to buildings and structures for which secondary effects need to be taken into consideration (TC-2(e, i)).

5.1. LOADS FROM TSUNAMIS

Selected observations concerning the treatment of tsunami loading conditions include the following:

(a) Hydrodynamic loads are a function of momentum flux, not only flow velocity or water height. The momentum flux per unit mass per unit width of the structure element (ΔM) is defined as follows:

$$\Delta M = hV^2 \quad (1)$$

where h is the fluid height and V is the flow velocity.

The momentum flux varies with time and location on the site. It should be noted that the maximum value ΔM_{max} over time does not equal $h_{max}V_{max}^2$, since maximum water depth and maximum flow velocity do not necessarily occur at the same time. The maximum momentum flux is the dependent loading parameter — not h_{max} or V_{max}^2 individually. The maximum value (ΔM_{max}) can be obtained by running detailed numerical simulation models or acquiring existing simulation data. If numerical models are not available, approximate formulas can be used to estimate the value.

(b) Simplified approaches, assumed to be conservative, are presented for a number of different cases (i.e. loading conditions). For example, the impulsive load can be defined as 1.5 times the hydrodynamic load, without further analyses being performed. A conservative definition of the combined hydrostatic load and hydrodynamic load can be used instead of treating the hydrodynamic load separately [13].

(c) The effective density of the fluid is a combination of water and suspended soil particles.

5.2. LOAD COMBINATIONS AND OTHER DESIGN CONSIDERATIONS

The tsunami protection system is designed on the basis of a combination of tsunami loading conditions and loading conditions associated with the plant's operational state. As discussed in the following subsections, the plant's operational state is assumed by considering factors such as the impact of the earthquake that generated the tsunami, the tsunami transit time and the effectiveness of the tsunami warning system.

5.2.1. Categorization of tsunamis

Tsunamis can be classified as local tsunamis or distant tsunamis according to their proximity to the site.

5.2.1.1. Local tsunami

A tsunami is called local when it is generated near the site. Local tsunamis can be generated by earthquakes, volcanic activity and landslides. In this publication, only earthquake induced tsunamis are considered. They represent the most frequent type of destructive tsunami.

5.2.1.2. Distant tsunami

Distant tsunamis are ocean wide tsunamis that arrive at places remote from their source after travelling across the ocean or sea basins.

5.2.1.3. Transit time

Transit time is the time required for the tsunami waves to travel from the source to the site.

5.2.2. Advance warning

The warning time from tsunami warning systems may differ significantly depending on the source of the tsunami.

Local tsunamis may have warning times that are extremely short. Therefore, decisions and actions taken to bring the NPP reactors to a hot shutdown condition need to be made in a timely manner if it is considered that there is not enough time for a cold shutdown. Similarly, any physical barriers not permanently in place need to be put in position as part of defence in depth provisions before the tsunami waves reach the site, but without endangering the plant's staff.

Distant tsunamis may have transit times (and warning times) of many hours, during which time the regulatory body, the licensee and their representatives (decision makers) can monitor the path and size of tsunami waves. Decisions are likely to be made to shut down the reactor and maintain it in a hot or cold shutdown state.

5.2.3. On-site effects

5.2.3.1. Local tsunami

By definition, local tsunamis are produced by earthquakes occurring close to the site. Consequently, it is expected that earthquake ground motion will be experienced at the site.

Depending on the earthquake source parameters that are likely to produce the tsunami (e.g. magnitude, distance and fault rupture characteristics), the immediate top level effects of shaking on the NPP site may be the following:

(a) Ground motion exceeding the DBE ground motion, causing no damage, minimal damage or significant damage to SSCs.
(b) Ground motion of amplitude lower than that of the DBE and greater than that of the operating basis earthquake, causing no damage or minimal damage to SSCs important to safety, but leading to automatic scram or manual shutdown of the operating NPP.
(c) Ground motion lower than that of the operating basis earthquake, but significant enough to cause automatic scram or manual shutdown.

5.2.3.2. Distant tsunami

By definition, distant tsunamis are produced by earthquakes at large distances from the NPP site. Therefore, no earthquake ground motion is experienced at the site.

5.2.4. Plant operational state at the time of the tsunami reaching the plant site

5.2.4.1. Local tsunami

For all levels of ground motion that has potential to generate a local tsunami, it is likely that the operating NPP will be shut down. The resulting plant operational state will be as follows:

(a) Minimal seismic damage has occurred; the plant is in hot shutdown.
(b) Minor seismic damage could have occurred to SSCs important to safety (including tsunami protection systems); the plant is in hot shutdown. The loading conditions from the earthquake ground motion at the DBE level need to be taken into account, in addition to the tsunami loading conditions.
(c) Loading conditions due to a beyond design basis earthquake have been experienced by the NPP prior to the tsunami's occurrence; the beyond design basis earthquake loading conditions need to be evaluated to determine the initial state of the plant prior to the tsunami occurring; alternative plant states are considered in the design process.
(d) For the purpose of the tsunami probabilistic safety assessment, the plant operational states described in (a), (b) or (c) above define the possible initial conditions. In addition, other possible initial conditions need to be evaluated; for example, loss of off-site power, restricted access to off-site or on-site emergency response personnel and equipment (such as the fire services, FLEX (diverse and flexible coping strategies) [11]) and

availability of evacuation routes. Also, the mission time used in the tsunami probabilistic safety assessment needs to be established in light of the need for on-site resources and off-site support of all types (personnel, and replenishment of materials, such as diesel fuel).

5.2.4.2. Distant tsunami

Owing to the advanced warning time and the information that has been made available to the decision makers, the plant operational state can reasonably be assumed to be in hot or cold shutdown.

5.2.5. Use of severe accident prevention and management approaches

In the evaluation of BDBEEs for the earthquake ground motion and the tsunami effects, severe accident prevention and management approaches may be implemented. However, the beyond design basis earthquake ground motion and the beyond design basis tsunami for tsunami hazards may need to be considered sequentially for the local tsunami. The effectiveness of these measures is dependent on the conditions as given in Section 2.3.3.

5.2.6. Load combinations

Load combinations can be established on the basis of normal operating conditions at the assumed alternative plant state(s), including the initial conditions as discussed above.

6. DESIGN AND QUALIFICATION OF MECHANICAL AND ELECTRICAL EQUIPMENT AND DISTRIBUTION SYSTEMS

Systems, components (mechanical), equipment (mechanical and electrical) and distribution systems (e.g. piping, cable trays/conduit, cables, heating ducts, ventilation and air-conditioning system, tubing) (SCE-DS)[3] that are important to safety are addressed in this section.

---

[3] SCE-DS stands for systems, components, equipment and distribution systems.

Two cases are distinguished: (a) items housed inside buildings and (b) items located outside the buildings and structures (i.e. in the yard). This distinction is important because the tsunami loading conditions are significantly different depending on the item's location. Items located inside buildings are primarily susceptible to inundation type loading conditions. Items in the yard are subject to the loading conditions on structures that are presented in Section 5.

6.1. TSUNAMI LOADING CONDITIONS INSIDE BUILDINGS

Items located inside buildings may be impacted by water penetration through openings in walls, roofs or other portions of the structure housing them, resulting in spray and inundation effects. Typical SCE-DS items inside buildings and susceptible to inundation failure modes are the following:

(a) Passive components: Penetrations comprising building watertight boundaries or building interior watertight boundaries.
(b) Active components: Components whose operation is necessary to perform the safety functions required, such as pumps, pump drive turbines, electrical motors, fans, refrigerating machines, emergency diesel generators, control air compressors, reciprocating pumps, valves and dampers. These include the electrical and instrumentation components that serve as auxiliary equipment for active components.
(c) Electrical and instrumentation components: Components that are necessary for electrical functions in order to perform the required safety functions, such as panels, devices, apparatuses and electrical circuits. These components are described below:
    (i) Panels: Assemblies that integrate structures made of materials such as steel, with internal components consisting of circuits, devices, apparatuses, electrical wires, cables and other elements. These are designed to perform functions such as switching and electrical power conversion. Panels also support control and operation systems (e.g. central control panels, local operation panels).
    (ii) Devices: Electrical and instrumentation components whose purpose is to convert electrical power or energy (e.g. transformers, storage batteries).
    (iii) Apparatuses: Elements that produce functions on electrical systems and instrument systems and perform operations for detecting, converting, mathematical operations, control, etc., in response to signals or electrical power handled in electrical and instrumentation components (e.g. various types of detector, transmitters).

(iv) Cable runs: Structures and electrical circuits composed of electrical wires, cables, conductors, etc. They are placed on supporting and protective structures that comprise other materials (e.g. cable trays, conduit pipes).

For TC-1 buildings that house SSCs important to safety, the tsunami design philosophy includes protection by a permanent tsunami protection system, such as a sea wall (TC-1(e)) and watertight doors, penetrations or other accesses (TC-1(i)), that prevent fluid (water plus sediment) from flowing.

If operator action is needed to implement an element of the tsunami protection system, procedures need to be in place and training implemented to provide high confidence that operator actions will occur. For example, if one or more accesses to a TC-1 building are open owing to maintenance or other actions, there is a need for verification of operator actions to close the accesses within the appropriate time frame.

Therefore, the issue for design and qualification of SCE-DS is to ensure that the TC-1 items are designed for the design basis tsunami, constructed to the design requirements and maintained during the plant life.

For SCE-DS items, a what-if analysis needs to be performed to verify that even if the TC-1(e) and TC-1(i) tsunami protection system elements fail, the SCE-DS will achieve their functional performance level for the design basis tsunami and the beyond design basis tsunami (see para. 5.35 of SSG-68 [3]).

6.2. TSUNAMI LOADING CONDITIONS OUTSIDE BUILDINGS

All loading conditions identified in Sections 4 and 5 apply to SCE-DS located in the yard. Examples of such SCE-DS are the following:

(a) Pumps;
(b) Tanks;
(c) Piping systems;
(d) Underground chases and contents;
(e) Measuring instruments (e.g. pressure gauges, flow meters), terminal boxes and other auxiliary equipment installed in electrical power panels (metal clad components), control panels, piping, etc.

Important loading conditions for yard located equipment are inundation, hydrostatic loads, buoyancy, hydrodynamic loads, impulsive loads, impact loads from debris or yard located items that become waterborne missiles, scouring, deposition tidal bore and drawdown.

6.3. SUPPLEMENTAL EVALUATIONS INSIDE AND OUTSIDE BUILDINGS

In-plant evaluations (walkdowns) are a necessary part of the tsunami protection design and evaluation. The term 'design' for existing plants includes upgrades and retrofit designs. For new plants, the term refers to the design of the tsunami protection system.

For existing plants, the procedure includes the following:

(a) In-office preparation, which includes review drawings, technical specifications, system descriptions, design requirements for leak tightness, and flood evaluations; documentation and planning of in-plant walkdowns.
(b) In-plant walkdowns, which cover inventory, evaluation and documentation of the as-is condition of all identified potential flow paths into TC-1 buildings and other structures (e.g. underground chases), as identified during in-office or in-plant walkdowns. Examples are accesses such as personnel and equipment access points, penetrations, underground penetrations to buildings for distribution systems (piping, cable, tubing, ducts). Documentation is needed of existing conditions and evaluation of whether they are leaktight. Finally, coordination with responsible plant personnel for external and internal flood evaluations is desirable.
(c) Assessment of whether tsunami protection system components are adequate to protect SSCs important to safety. If they are not, provision needs to be made of design barriers or functional resistance measures to meet tsunami design and evaluation acceptance criteria.

For new plants being designed, but not constructed, the procedures in Sections 3–6 may be followed, then supplemented when construction is complete by in-plant verifications of the elements of the tsunami protection system.

7. DESIGN TO ADDRESS SPECIAL ISSUES OF TSUNAMIS

This section covers the following special issues that are not encountered in usual design situations:

— Soil failure modes;
— Intake structure blocking and drawdown;
— Tsunami induced fire.

7.1. SOIL FAILURE MODES

Soil failures can have a significant effect on TC-1 and TC-2 SSCs important to safety, if these SSCs are founded on soil media susceptible to soil related failures from ground shaking alone and from tsunami related phenomena (e.g. scouring) that could also cause such failures.

For local tsunamis, it is particularly important to assess the effects of the local earthquake ground motion that the local tsunami is assumed to cause on the installation and site. This sets the initial conditions for the plant state (see Sections 2.3, 2.5, 5.3.1 and 5.3.3).

Failure modes of interest are the following:

(a) Liquefaction (including lateral spreading): This phenomenon is of particular interest for the combined set of seismic category 1 [9] and TC-1 NPP SSCs. In addition, geotechnical structures or elements (e.g. berm, channel) and selected mechanical and/or electrical components (e.g. tanks) may be susceptible to liquefaction, including lateral spreading.

(b) Seismically induced land sliding: This phenomenon is of particular interest for slopes present on the site, either natural or human made. The result of the landslide could be a direct or indirect effect on seismic category 1 or TC-1 NPP SSCs.

Soil failure modes need to be evaluated by senior, experienced geotechnical engineers.

7.2. INTAKE STRUCTURE BLOCKING AND DRAWDOWN

The structural integrity of the intake structure and the structural integrity and operability of the SSCs housed therein represent a special case. Design loading conditions include the following:

(a) Significant foundation/structure movement due to tsunami loading conditions of hydrostatic loads (including buoyancy), hydrodynamic loads, impulsive loads and debris impact loads. Scouring can also play a role in structural integrity.

(b) Significant tsunami loading conditions on systems and components are hydrostatic loads, deposition and drawdown.

Design considerations for the intake structure and SSCs housed therein are the following:

(a) There is assurance that no foundation and/or structural failure will occur from the design basis tsunami and is highly unlikely from the beyond design basis tsunami.
(b) The design is such that the water intake opening of seawater pumps is not blocked by deposition of sand (sand drift). If deposition of sand near the water intake opening is a necessary design condition, the structure has provisions to make it difficult for sand to flow into the water intake opening.
(c) Seawater pumps are designed so that the pump is not damaged and is available to perform its function if a mixture of sand in sea water is assumed to ingress into the structure owing to the design basis tsunami. By implication, the seawater pumps and supporting systems are classified as TC-1(t). Protection at some level is provided by travelling screens in this area.

Examples of damage modes assumed are the following:

(a) Damage due to inflow of sand into gaps (e.g. abrasion of submerged bearings);
(b) Damage due to inflow of sand into the main pump unit;
(c) Damage to other parts (e.g. damage to parts handling drift sand).

In the tsunami resistant design of seawater pumps, the assessment conditions are the concentration of sand by grain size within the water intake facility, the tsunami duration and the operating time needed during and after tsunami occurrence. Using an appropriate method, it is confirmed that the seawater pumps are not damaged and that the water intake function is maintained.

Water intake facilities and seawater pumps are designed so that the inflow of sea water from the water intake opening can be continuously maintained and the water intake of the TC-1 seawater pumps can be maintained in response to the drawdown assumed during a tsunami.

Designs providing a structure to retain sea water temporarily in a water intake facility may be implemented so that the seawater pumps are able to draw water as necessary even if drawdown due to the tsunami has occurred.

If none of these options is feasible, a design is implemented in which the height of the seawater surface does not go below the level at which water is able to be drawn by the TC-1 seawater system pumps, even if drawdown occurs, or alternative backup systems are available to provide the necessary functionality of these pumps (i.e. cooling).

Even if the water level drops below the level assumed for the design basis tsunami and the seawater pumps temporarily stop, there is a possibility that they can be restarted as the water level recovers. In this case, it is necessary to confirm that no permanent damage to the intake structure and seawater pumps has occurred.

The general design procedures to address drawdown, also known as backwash, are shown in Fig. 3.

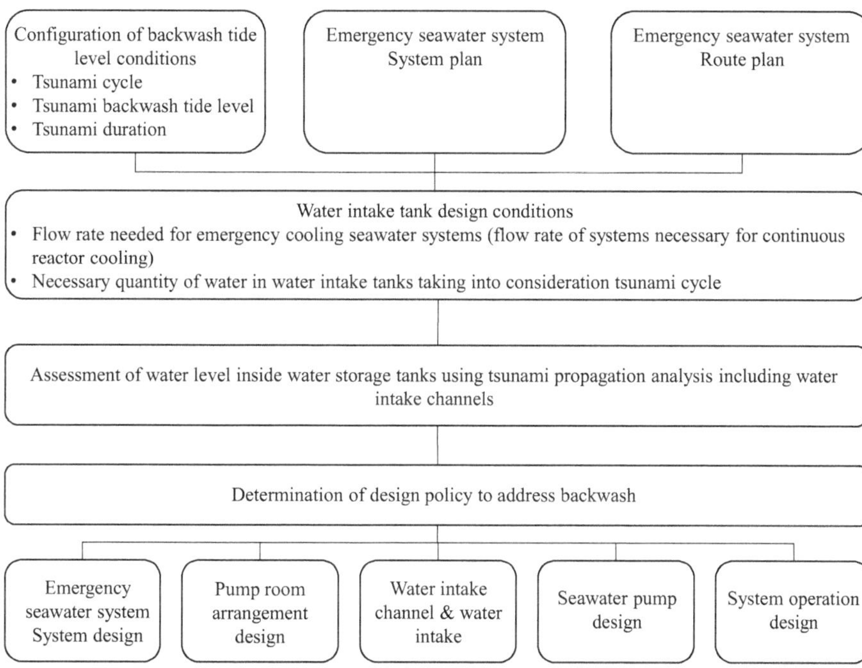

FIG. 3. Design procedures to address backwash.

7.3. TSUNAMI INDUCED FIRE

The specific measures for protection against tsunami caused fires affecting an NPP are as follows:

(a) Preventing fires from occurring: A tsunami design for equipment containing combustible material (categorized as TC-2; e.g. outdoor oil tanks) is implemented, which implies no failure that would release combustible material. Moreover, a tsunami design for electrical equipment that may become an ignition source (TC-2) is implemented.

(b) Detecting fires and extinguishing fires: Fire foam, portable fire pumps and other firefighting equipment are made available for deployment near SSCs important to safety, with the goal of protecting them from tsunami induced fire. The overall objective is that firefighting activities can proceed expeditiously if such a fire occurs.

(c) Mitigating impact of fires: SSCs important to safety are protected from fire damage by installing barriers (facilities for countering flotsam) to prevent the approach of flotsam or by implementing fire prevention measures on the surface of the SSCs (building walls, penetrations, doors, etc.). However, this does not apply if the equipment containing combustible material is located at a considerable distance, if the tsunami's directionality makes the risk of a tsunami induced fire negligible, or if the site is a dry site with a negligible risk of such a fire.

(d) Measures concerning operational aspects include the option of an on-site, permanently stationed firefighting service. Workers and other personnel monitor and patrol within the site, and the firefighting service promptly fights any fires.

8. TSUNAMI DETECTION AND OPERATION

Advance warning of the potential for a tsunami to impact a nuclear installation site is an important element in protecting the installation from damage due to a tsunami. In general, actions to be taken before the arrival of the tsunami wave are focused on putting in place physical barriers that are not permanently in place and in shutting down the installation. In the case of an NPP, this includes bringing it into hot or cold shutdown depending on the estimated time of arrival of the tsunami.

8.1. IAEA GUIDANCE

IAEA guidance emphasizes the importance of an active, operating tsunami warning system. Paragraph 9.12 of SSG-18 [2] states:

"In regions where there is no local, national or regional tsunami warning system in place, the operating organization should receive messages from the national, regional or global seismic monitoring centre to be informed of occurrences of major earthquakes."

Paragraph 9.14 of SSG-18 [2] states that "In coastal regions without sea level monitoring stations, a real time sea level monitoring network should be set up for the collection and real time transmission of data to the nuclear installation."

It is necessary to establish contacts with existing tsunami warning and watch centres at the international, national, interregional, regional and local levels, if they exist. The intent is for the nuclear installation and its management to receive alerts and messages related to local and distant tsunamis in a timely manner. Detailed information is provided in annex III of SSG-18 [2].

Reference [14] highlights the need to ensure that, in considering external natural hazards, an active tsunami warning system is established with the provision for immediate operator action. It also indicates that severe accident management guidelines and associated procedures need to take into account the potential unavailability of instruments, lighting and power, as well as abnormal conditions, including abnormal plant behaviour and high radiation fields.

8.2. OPERATION IN RELATION TO A TSUNAMI WARNING SYSTEM

Aspects of the operational process that may be influenced by an effective tsunami warning system include the following:

(a) The tsunami protection system may be assumed to be in place and operational if operating procedures for addressing tsunamis are developed and implemented, and training is provided (and refreshed on a regular basis) on the actions to be taken if the plant staff has been informed of the potential arrival of a tsunami. One important element of the operating procedures is to ensure that physical barriers (permanent and non-permanent) are in place when the tsunami strikes. Examples of non-permanent barriers include sea walls with accesses open for the inflow and outflow of water under normal operating conditions, and watertight doors in buildings important to safety that can be closed and secured.
(b) Load combinations may be affected by the assumption that the plant state has changed from power operation to a shutdown condition.

9. MANAGEMENT SYSTEM FOR TSUNAMI DESIGN AND EVALUATION

9.1. APPLICATION OF THE MANAGEMENT SYSTEM

Designing and operating a nuclear installation to resist the effects of a tsunami require a multidisciplinary effort that includes numerous interfaces and coordinated activities:

(a) A management system applicable to all organizations involved in the tsunami design and evaluation is established and implemented before the start of the tsunami design and safety evaluation programme. Recommendations on the application of the management system for hazard assessments are provided in section 11 of SSG-18 [2].
(b) The management system covers all processes and activities of the programme for tsunami design and tsunami safety evaluation; in particular, those relating to data collection and data processing, field and laboratory investigations, analyses of tsunami wave propagation from source to control point, analyses of the impacts of tsunami phenomena onto the site, and design and evaluations of the nuclear installation regarding tsunami protection. It also covers those processes and activities corresponding to the re-evaluation phase of the programme.
(c) All interfaces between disciplines, organizations and evaluations, including data exchanges are established.
(d) Tsunami specific operational procedures are developed and reviewed, and relevant training is given.
(e) Participatory peer review is implemented, as specified below.
(f) Documentation is developed and maintained under the organization's quality assurance programme defined in the management system.

9.2. PARTICIPATORY PEER REVIEW

A participatory peer review of the implementation of the tsunami design and evaluation process is performed. 'Participatory' means that the peer review team is involved in all stages of the design and evaluation. The team is independent of the work being performed.

The peer review is conducted by experts in the areas of seismic hazard, tsunami hazard, earthquake engineering (geotechnical, civil, structural, mechanical, electrical, flood), safety and systems engineering. At least one member of the participatory peer review team needs to have operations experience.

The following peer reviews are performed at different stages in the evaluation process:

(a) Seismic and tsunami hazard assessments, including tsunami phenomena propagated onto the site;
(b) A systems and operations review to verify the identification of SSCs important to safety and their locations, including collaboration with engineering to identify buildings and structures to be protected;
(c) Geotechnical and civil engineering review of preliminary selection of the elements of the tsunami protection system;
(d) Geotechnical, civil and structural engineering review of tsunami loading conditions for design and evaluations regarding the design basis tsunami and beyond design basis tsunami;
(e) Engineering review of all elements of the tsunami protection system;
(f) Review of the in-plant evaluations (walkdowns);
(g) Review of operational procedures related to tsunami protection.

The findings of the peer reviews are documented in a report, and the recommendations are considered for implementation at the nuclear installation.

9.3. DOCUMENTATION AND RECORDS

Design documentation is retained as required by the quality assurance programme defined in the management system. This, and other documentation on the evaluation of the nuclear installation when subjected to the beyond design basis tsunami, are retained for review and future application. Typical documentation of the results of the tsunami design and evaluation is a report containing the following:

(a) Summary of the characteristics of the design basis tsunami and the beyond design basis tsunami, including governing earthquake source(s) and characteristics, tsunami phenomena at control point, propagation results of the tsunami at the nuclear site, and tsunami loading conditions imposed on TC-1 and TC-2 items;
(b) Table of TC-1 and TC-2 tsunami protection system elements and all item specific design loading conditions;
(c) Statements affirming no cliff edge effects for the design basis tsunami and, preferably, the beyond design basis tsunami;

(d) Summary of operational guidance or requirements for the tsunami;
(e) Operator actions required and the evaluation of their likelihoods of success;
(f) Summary of the in-plant evaluation (walkdown report) summarizing findings and plant-wide observations, if any;
(g) Confirmation of the satisfactory performance of the containment and containment system when subjected to the tsunami phenomena and loading conditions, as well as the high confidence of low probability of failure or fragility functions (if required);
(h) Planned use of non-permanent equipment;
(i) Peer review reports.

Specific plant procedures are prepared for dealing with the required response actions before, during and after a tsunami.

9.4. CONFIGURATION CONTROL

The operator of a nuclear installation implements a configuration control programme to ensure that in the future the design and construction of new SSCs, or modifications to existing SSCs, do not invalidate the effectiveness of TC-1 and TC-2 elements.

The operator implements operating procedures in all plant states to ensure that the tsunami protection system continues to be effective when maintenance, refuelling or other activities are implemented at the nuclear installation.

REFERENCES

[1] INTERNATIONAL ATOMIC ENERGY AGENCY, Safety of Nuclear Power Plants: Design, IAEA Safety Standards Series No. SSR-2/1 (Rev. 1), IAEA, Vienna (2016).

[2] INTERNATIONAL ATOMIC ENERGY AGENCY, WORLD METEOROLOGICAL ORGANIZATION, Meteorological and Hydrological Hazards in Site Evaluation for Nuclear Installations, IAEA Safety Standards Series No. SSG-18, IAEA, Vienna (2011).

[3] INTERNATIONAL ATOMIC ENERGY AGENCY, Design of Nuclear Installations Against External Events Excluding Earthquakes, IAEA Safety Standards Series No. SSG-68, IAEA, Vienna (2021).

[4] NUCLEAR REGULATORY COMMISSION, Regulatory Guide 1.174, Revision 3, An Approach for Using Probabilistic Risk Assessment in Risk-Informed Decisions on Plant-Specific Changes to the Licencing Basis, NRC, Washington, DC (2018).

[5] NUCLEAR REGULATORY COMMISSION, "Probabilistic risk assessment and severe accident evaluation for new reactors", Chapter 19, Section 19.0, Standard Review Plan for the Review of Safety Analysis Reports for Nuclear Power Plants: LWR Edition, NUREG-0800, NRC, Washington, DC (2015).

[6] CANADIAN NUCLEAR SAFETY COMMISSION, Design of Reactor Facilities, Version 2.1, REGDOC-2.5.2, CNSC, Ottawa (2023).

[7] NUCLEAR REGULATION AUTHORITY JAPAN, Enforcement of the New Regulatory Requirements for Commercial Nuclear Power Reactors, NRA Japan, Tokyo (2013),
https://www.nra.go.jp/data/000067212.pdf

[8] INTERNATIONAL ATOMIC ENERGY AGENCY, Considerations on the Application of the IAEA Safety Requirements for the Design of Nuclear Power Plants, IAEA-TECDOC-1791, IAEA, Vienna (2016).

[9] INTERNATIONAL ATOMIC ENERGY AGENCY, Seismic Design for Nuclear Installations, IAEA Safety Standards Series No. SSG-67, IAEA, Vienna (2021).

[10] UNITED STATES NUCLEAR REGULATORY COMMISSION, Japan Lessons-Learned Project Directorate, Compliance with Order EA-12-049, Order Modifying Licenses with Regard to Requirements for Mitigation Strategies for Beyond-Design-Basis External Events, Interim Staff Guidance JLD-ISG-2012-01 (Rev. 0), NRC, Washington, DC (2012).

[11] UNITED STATES NUCLEAR ENERGY INSTITUTE, Diverse and Flexible Coping Strategies (FLEX) Implementation Guide, Rep. NEI 12-06 (Rev. 0), NEI, Washington, DC (2012).

[12] NUCLEAR STANDARDS COMMITTEE OF JAPAN ELECTRIC ASSOCIATION, Technical Code for Tsunami Design of Nuclear Power Plants, Rep. JEAC 4629–2021, Japan Electric Association, Tokyo (2021).

[13] FIRE AND DISASTER MANAGEMENT AGENCY, JAPAN MINISTRY OF INTERNAL AFFAIRS AND COMMUNICATIONS, Report of Investigation and Examination on Tsunami and Inundation Measures at Dangerous Facilities, MIC, Tokyo (2009),
https://www.fdma.go.jp/pressrelease/houdou/items/h21/2105/210526-1houdou/02_stunami_houkokusyo.pdf

[14] INTERNATIONAL ATOMIC ENERGY AGENCY, The Great East Japan Earthquake Expert Mission, IAEA International Fact Finding Expert Mission of the Fukushima Dai-ichi NPP Accident Following the Great East Japan Earthquake and Tsunami, Report to the IAEA Member States, IAEA, Vienna (2011).

Annex I

REGULATIONS AND GUIDELINES IN THE UNITED STATES OF AMERICA

This annex contains an extensive list of United States publications (regulations, guidelines, etc.) related to seismic topics, including responses to the Fukushima Daiichi accident due to seismic and tsunami phenomena.

I–1. NUCLEAR REGULATORY COMMISSION REGULATIONS

(1) NUCLEAR REGULATORY COMMISSION, "Appendix A to Part 50 — General Design Criteria for Nuclear Power Plants", Domestic Licensing of Production and Utilization Facilities, 10 CFR 50, NRC, Washington, DC (2022).
(2) NUCLEAR REGULATORY COMMISSION, "Appendix S to Part 50 — Earthquake Engineering Criteria for Nuclear Power Plants", Domestic Licensing of Production and Utilization Facilities, 10 CFR 50, NRC, Washington, DC (2022); para. IV(c).
(3) NUCLEAR REGULATORY COMMISSION, Licenses, Certifications, and Approvals for Nuclear Power Plants, 10 CFR 52, NRC, Washington, DC (2022); paras 52.17(a)(vi) and 52.79 (a)(1)(iii).
(4) NUCLEAR REGULATORY COMMISSION, Reactor Site Criteria, 10 CFR 100, NRC, Washington, DC (2022); paras 100.10(c), 100.20(c), 100.23(d)(4) and Appendix A.

I–2. NUCLEAR REGULATORY COMMISSION GUIDANCE

I–2.1. Nuclear Regulatory Commission Standard Review Plan

(5) NUCLEAR REGULATORY COMMISSION, "Flooding protection requirements", Rev. 3, Standard Review Plan for the Review of Safety Analysis Reports for Nuclear Power Plants: LWR Edition, Rep. NUREG-0800, NRC, Washington, DC (2007) Section 2.4.10.
(6) NUCLEAR REGULATORY COMMISSION, "Floods", Rev. 4, Standard Review Plan for the Review of Safety Analysis Reports for Nuclear Power

Plants: LWR Edition, Rep. NUREG-0800, NRC, Washington, DC (2007) Section 2.4.2.
(7) NUCLEAR REGULATORY COMMISSION, "Tsunami hazards", Rev. 4, Standard Review Plan for the Review of Safety Analysis Reports for Nuclear Power Plants: LWR Edition, Rep. NUREG-0800, NRC, Washington, DC (2018) Section 2.4.6.
(8) NUCLEAR REGULATORY COMMISSION, "Ultimate heat sink", Rev. 3, Standard Review Plan for the Review of Safety Analysis Reports for Nuclear Power Plants: LWR Edition, Rep. NUREG-0800, NRC, Washington, DC (2007) Section 9.2.5.
(9) NUCLEAR REGULATORY COMMISSION, "Internal flood protection for onsite equipment failures", Rev. 3, Standard Review Plan for the Review of Safety Analysis Reports for Nuclear Power Plants: LWR Edition, Rep. NUREG-0800, NRC, Washington, DC (2007) Section 3.4.1.
(10) NUCLEAR REGULATORY COMMISSION, "Analysis procedures", Rev. 3, Standard Review Plan for the Review of Safety Analysis Reports for Nuclear Power Plants: LWR Edition, Rep. NUREG-0800, NRC, Washington, DC (2007) Section 3.4.2.
(11) NUCLEAR REGULATORY COMMISSION, Tsunami Hazard Assessment at Nuclear Power Plant Sites in the United States of America, Rep. NUREG/CR-6966, NRC, Washington, DC (2009).
(12) NUCLEAR REGULATORY COMMISSION, "Probable maximum surge and seiche flooding", Rev. 3, Standard Review Plan for the Review of Safety Analysis Reports for Nuclear Power Plants: LWR Edition, Rep. NUREG-0800, NRC, Washington, DC (2007) Section 2.4.5.
(13) NUCLEAR REGULATORY COMMISSION, "Probable maximum flood on streams and rivers", Rev. 3, Standard Review Plan for the Review of Safety Analysis Reports for Nuclear Power Plants: LWR Edition, Rep. NUREG-0800, NRC, Washington, DC (2007) Section 2.4.3.
(14) NUCLEAR REGULATORY COMMISSION, "Potential dam failures", Rev. 3, Standard Review Plan for the Review of Safety Analysis Reports for Nuclear Power Plants: LWR Edition, Rep. NUREG-0800, NRC, Washington, DC (2007) Section 2.4.4.

1–2.2. Nuclear Regulatory Commission Regulatory Guides

(15) NUCLEAR REGULATORY COMMISSION, Ultimate Heat Sink for Nuclear Power Plants (Rev. 2), Regulatory Guide 1.27, NRC, Washington, DC (1976).

(16) NUCLEAR REGULATORY COMMISSION, Seismic Design Classification (Rev. 4), Regulatory Guide 1.29, NRC, Washington, DC (2007).
(17) NUCLEAR REGULATORY COMMISSION, Flood Protection for Nuclear Power Plants (Rev. 1), Regulatory Guide 1.102, NRC, Washington, DC (1976).
(18) NUCLEAR REGULATORY COMMISSION, Inspection of Water-Control Structures Associated with Nuclear Power Plants (Rev. 1), Regulatory Guide 1.127, NRC, Washington, DC (1978).
(19) NUCLEAR REGULATORY COMMISSION, Design Basis Floods for Nuclear Power Plants (Rev. 2), Regulatory Guide 1.59, NRC, Washington, DC (1977).

1–2.3. Nuclear Regulatory Commission and Electric Power Research Institute Regulatory and Guidance Documents Post-Fukushima Accident (Near Term Task Force)

(20) NUCLEAR REGULATORY COMMISSION, Recommendations for Enhancing Reactor Safety in the 21st Century: Near-Term Task Force Review of Insights from the Fukushima Dai-ichi Accident, Commission Paper SECY-11-0093 (ML11186A950), NRC, Washington, DC (2011).
(21) NUCLEAR REGULATORY COMMISSION, Prioritization of Recommended Actions to be Taken in Response to Fukushima Lessons Learned, Commission Paper SECY-11-0137, NRC, Washington, DC (2011).
(22) NUCLEAR REGULATORY COMMISSION, Recommended Actions to be Taken Without Delay from the Near-Term Task Force Report, Commission Paper SECY-11-0124, NRC, Washington, DC (2011).
(23) ELECTRIC POWER RESEARCH INSTITUTE, Fleet Seismic Core Damage Frequency Estimates for Central and Eastern U.S. Nuclear Power Plants Using New Site-Specific Seismic Hazard Estimates, EPRI, Palo Alto, CA (2014).
(24) NUCLEAR REGULATORY COMMISSION, Request for Information Pursuant to Title 10 of the Code of Federal Regulations 50.54(f) Regarding Recommendations 2.1, 2.3, and 9.3, of the Near-Term Task Force Review of Insights from the Fukushima Dai-ichi Accident, ML12053A340, NRC, Washington, DC (2012).
(25) NUCLEAR REGULATORY COMMISSION, GI-199 Established, ML051600272, NRC, Washington, DC (2005).
(26) ELECTRIC POWER RESEARCH INSTITUTE, CEUS Ground Motion Project Final Report, Rep. EPRI 1009684, EPRI, Palo Alto, CA (2004).

(27) NUCLEAR REGULATORY COMMISSION, Resolution of Generic Safety Issues: Issue 199: Implications of Updated Probabilistic Seismic Hazard Estimates in Central and Eastern United States on Existing Plants (NUREG-0933, Main Report with Supplements 1–35), ML101970221, NRC, Washington, DC (2010).
(28) NUCLEAR REGULATORY COMMISSION, UNITED STATES DEPARTMENT OF ENERGY, ELECTRIC POWER RESEARCH INSTITUTE, Central and Eastern United States Seismic Source Characterization for Nuclear Facilities, EPRI, Palo Alto, CA (2012).
(29) ELECTRIC POWER RESEARCH INSTITUTE, Seismic Walkdown Guidance for Resolution of Fukushima Near Term Task Force Recommendation 2.3: Seismic, Rep. EPRI 1025286, EPRI, Palo Alto, CA (2012).
(30) ELECTRIC POWER RESEARCH INSTITUTE, Seismic Evaluation Guidance: Screening, Prioritization and Implementation Details (SPID) for the Resolution of Fukushima Near-Term Task Force Recommendation 2.1: Seismic, Rep. EPRI 1025287, EPRI, Palo Alto, CA (2012).
(31) NUCLEAR REGULATORY COMMISSION, Screening and Prioritization Results Regarding Information Pursuant to Title 10 of the Code of Federal Regulations 50.54(f) Regarding Seismic Hazard Re-evaluations for Recommendation 2.1 of the Near-Term Task Force Review of Insights from the Fukushima Dai-ichi Accident, ML14111A147, NRC, Washington, DC (2014).
(32) NUCLEAR REGULATORY COMMISSION, Guidance on Performing a Seismic Margin Assessment in Response to the March 2012 Request for Information Letter, JLD-ISG-2012-04, NRC, Washington, DC (2012).
(33) ELECTRIC POWER RESEARCH INSTITUTE, Seismic Evaluation Guidance, Augmented Approach for the Resolution of Fukushima Near-Term Task Force Recommendation 2.1: Seismic, Rep. EPRI 3002000704, EPRI, Palo Alto, CA (2013).
(34) NUCLEAR REGULATORY COMMISSION, Design Basis Flood Estimation for Site Characterization at Nuclear Power Plants in the United States of America, Rep. NUREG/CR-7046, NRC Washington, DC (2011).
(35) NUCLEAR REGULATORY COMMISSION, Guidance for Performing a Tsunami, Surge, or Seiche Hazard Assessment, Japan Lessons-Learned Project Directorate Interim Staff Guidance, Rep. JLD-ISG-2012-06, NRC, Washington, DC (2012).
(36) NUCLEAR REGULATORY COMMISSION, Performance of an Integrated Assessment for Flooding, Japan Lessons-Learned Project Directorate Interim Staff Guidance, Rep. JLD-ISG-2012-05, NRC, Washington, DC (2012).

(37) NUCLEAR REGULATORY COMMISSION, Research Activities FY 2012– FY 2014, Rev. 2, Rep. NUREG-1925, NRC, Washington, DC (2013).

1–2.4. Nuclear Regulatory Commission Interim Staff Guidance Associated with Japan Lessons Learned

(38) NUCLEAR REGULATORY COMMISSION, Compliance with Order EA-12-049, Order Modifying Licenses with Regard to Requirements for Mitigation Strategies for Beyond-Design-Basis External Events, Rep. JLD-ISG-2012-01 (Rev. 0), NRC, Washington, DC (2012).
(39) NUCLEAR REGULATORY COMMISSION, Compliance with Order EA-12-050, Reliable Hardened Containment Vents, Rep. JLD-ISG-2012-02 (Rev. 0), NRC, Washington, DC (2012).
(40) NUCLEAR REGULATORY COMMISSION, Compliance with Order EA-12-051, Reliable Spent Fuel Pool Instrumentation, Rep. JLD-ISG-2012-03 (Rev. 0), NRC, Washington, DC (2012).
(41) NUCLEAR REGULATORY COMMISSION, Guidance for Performing the Integrated Assessment for Flooding, Rep. JLD-ISG-2012-05 (Rev. 0), NRC, Washington, DC (2012).
(42) NUCLEAR REGULATORY COMMISSION, Guidance for Performing a Tsunami, Surge, or Seiche Hazard Assessment, Rep. JLD-ISG-2012-06 (Rev. 0), NRC, Washington, DC (2013).
(43) NUCLEAR REGULATORY COMMISSION, Guidance for Assessment of Flooding Hazards Due to Dam Failure, Rep. JLD-ISG-2013-01 (Rev. 0), NRC, Washington, DC (2013).

1–3. OTHER STANDARDS, GUIDELINES AND DOCUMENTS

(44) UNITED STATES FEDERAL EMERGENCY MANAGEMENT AGENCY, Guidelines for Design of Structures for Vertical Evacuation from Tsunamis, 2nd edn, Rep. FEMA P646, FEMA, Washington, DC (2012).
(45) AMERICAN CONCRETE INSTITUTE, ASME Boiler and Pressure Vessel Code: Code for Concrete Containments, Section III, Div. 2, Standard ACI 359, ACI, Farmington Hills, MI (2007).
(46) AMERICAN CONCRETE INSTITUTE, Code Requirements for Nuclear Safety Related Concrete Structures, Standard ACI 349, ACI, Farmington Hills, MI (2007).
(47) AMERICAN INSTITUTE OF STEEL CONSTRUCTION, Specifications for the Design, Fabrication and Erection of Structural Steel for Buildings, 8th edn, AISC, New York (1980).

(48) AMERICAN INSTITUTE OF STEEL CONSTRUCTION, Specification for Safety-Related Structures for Design, Fabrication and Erection of Steel Safety Related Structures for Nuclear Facilities, Standard ANSI/AISC N-690 Including Supplement 2, ANSI, Washington, DC (1994).
(49) AMERICAN SOCIETY OF CIVIL ENGINEERS, "Tsunami loads and effects", Minimum Design Loads for Buildings and Other Structures, ASCE/SEI 7-16, ASCE, Reston, VA (2016) Ch. 6.
(50) UNITED STATES ARMY CORPS OF ENGINEERS, Coastal Engineering Manual, Rep. EM 1110-2-1100, USACE, Washington, DC (2008).
(51) AMERICAN INSTITUTE OF STEEL CONSTRUCTION, Specifications for the Design, Fabrication and Erection of Structural Steel for Buildings — 1946, AISC, Chicago, IL (2020).
(52) AMERICAN SOCIETY OF MECHANICAL ENGINEERS, Rules for Construction of Nuclear Power Plant Components, Section III, Div. I, including Addenda, ASME, New York (2004).
(53) INSTITUTE OF ELECTRICAL AND ELECTRONICS ENGINEERS, IEEE Recommended Practice for Seismic Qualification of Class 1E Equipment for Nuclear Power Generating Stations, IEEE Standard 344-2004, IEEE, Piscataway, NJ (2004).
(54) INSTITUTE OF ELECTRICAL AND ELECTRONICS ENGINEERS, Standard for Qualifying Class 1E Equipment for Nuclear Power Generating Stations, IEEE Standard 323-2003, IEEE, Piscataway, NJ (2003).
(55) NUCLEAR ENERGY INSTITUTE, Diverse and Flexible Coping Strategies (FLEX) Implementation Guide, Rep. NEI 12-06 (Rev. 0), NEI, Washington, DC (2012).
(56) NATIONAL OCEANIC AND ATMOSPHERIC ADMINISTRATION, Scientific and Technical Issues in Tsunami Hazard Assessment of Nuclear Power Plant Sites, NOAA Technical Memorandum OAR PMEL-136, Pacific Marine Environmental Laboratory, Seattle, WA (2007).
(57) TIO, K.K., SOMERVILLE, P., POLET, J., Probabilistic Tsunami Hazard in California, Rep. PEER 2010/108, University of California Pacific Earthquake Engineering Research Center, Berkeley, CA (2010).
(58) NUCLEAR ENERGY INSTITUTE, Guidelines for Performing Verification Walkdowns of Plant Flood Protection Features, Rep. NEI 12-07 (Rev. 0-A), NEI, Washington, DC (2012).
(59) NUCLEAR REGULATORY COMMISSION, Guidelines for Performing Verification Walkdowns of Plant Flood Protection Features, Endorsement of NEI 12-07, NRC, Washington, DC (2012).

Annex II

GUIDELINES FOR TSUNAMI HAZARDS AND DESIGN IN JAPAN

II–1. NUCLEAR REGULATION AUTHORITY REGULATIONS AND GUIDELINES

On the basis of the lessons learned from the accident at the Fukushima Daiichi nuclear power plant after the 2011 earthquake, the Nuclear Regulation Authority developed and enforced new regulations and guidelines in 2013 and a backfitting system has been introduced for existing nuclear power plants. In addition, even if an accident or natural disaster that exceeds design basis assumptions occurs in Japan, measures are required to prevent core damage, containment vessel damage and release of radioactive material. The overview of regulations and guidelines is summarized in Ref. [II–1].

Regulations and guidelines related to the derivation of the design basis tsunami are provided in Refs. [II–2 to II–4]. An overview is as follows. Design basis tsunamis are developed using a numerical simulation method, with uncertainties taken into consideration by selecting multiple generation sources (including not only earthquakes but also landslides, slope failure, etc.) and their combinations.

The runup heights resulting from design basis tsunamis have to exceed any tsunami/inundation heights estimated from observation data, including historical records and tsunami deposits, if they exist.

Regulations and guidelines related to the tsunami design for structures, systems and components are provided in Refs [II–2 to II–7]. An overview is as follows. Multilayered protective measures are taken to fulfil the requirement that the safety functions of structures, systems and components important to safety will not be compromised by a design basis tsunami. Installation of a sea wall to prevent site inundation and watertight doors to prevent the flooding of buildings are examples of multilayered protective measures.

II–2. ACADEMIC AND PRIVATE STANDARDS, GUIDELINES, CODES AND TECHNICAL REPORTS IN JAPAN

Japanese academic and private codes, guidelines and technical reports for designing and operating nuclear installations to resist the impact of tsunamis had been published before the 2011 earthquake but were updated, and new ones were added, since 2011 to incorporate lessons learned from the earthquake and tsunami and their impact. Reference [II–8] is a technical report published by the Japan Society of Civil Engineers in 2002 and addresses the methodology and numerical simulation technologies on the deterministic hazard analysis for tsunamis generated by earthquakes, which have been used to determine design basis tsunamis. After the 2011 earthquake, the Japan Society of Civil Engineers collected the latest knowledge on earthquakes and tsunamis and updated the technical report in 2016 [II–9]. A notable feature of Ref. [II–9] is the proposal of a new methodology for probabilistic tsunami hazard analysis. This methodology is based on a logic-tree approach, where epistemic and aleatory uncertainties are systematically taken into account in the assessment. In addition, a deterministic hazard analysis methodology for tsunamis generated by landslides, as well as earthquakes, is proposed. Furthermore, Ref. [II–9] includes methods and technologies for the evaluation of tsunami loads: prediction models and numerical simulation technologies for hydrostatic load; buoyancy; hydrodynamic load; debris impact loads on sea walls; breakwaters; buildings and tanks; and suspended sediment depositions. Evaluation methods for tsunami loads described in guidelines initiated by the Japanese Government for port facilities and tsunami evacuation buildings (Refs [II–10 to II–12]) are also incorporated in Ref. [II–9].

The Japan Electric Association published technical guidelines and codes in 2016 and 2021 [II–13 to II–16], which address the basic design concepts for tsunami protection facilities such as sea walls, prevention facilities from flooding into buildings such as watertight doors, and tsunami designs for outdoor tanks and piping.

The Atomic Energy Society of Japan published a tsunami probabilistic risk assessment standard in 2012 and updated it in 2016 [II–17 to II–18]. The standard describes the methodologies for the probabilistic tsunami hazard analysis, fragility analysis and accident sequence analysis against tsunamis. The standard also includes a method for determining probabilistically defined tsunami scenarios, which can be used for evaluations of tsunami impact for the design basis tsunami and the beyond design basis tsunami.

REFERENCES TO ANNEX II

[II–1] NUCLEAR REGULATION AUTHORITY, SARIS Summary Report, NRA, Tokyo (2015), https://www.nra.go.jp/data/000148577.pdf

[II–2] NUCLEAR REGULATION AUTHORITY, The NRA Ordinance on Standards for the Location, Structure, and Equipment of Commercial Power Reactors, NRA, Tokyo (2022) (in Japanese), https://elaws.e-gov.go.jp/document?lawid=425M60080000005

[II–3] NUCLEAR REGULATION AUTHORITY, The Regulatory Guide of NRA Ordinance on Standards for the Location, Structure, and Equipment of Commercial Power Reactors, NRA, Tokyo (2023) (in Japanese),
https://www.nra.go.jp/data/000382455.pdf
[II–4] NUCLEAR REGULATION AUTHORITY, Guidelines for Design Basis Tsunamis and Design Policy Against Tsunami, NRA, Tokyo (2021) (in Japanese),
https://www.nra.go.jp/data/000396219.pdf
[II–5] NUCLEAR REGULATION AUTHORITY, The NRA Ordinance on Technical Standards for Commercial Power Reactor Facilities, NRA, Tokyo (2022) (in Japanese),
https://elaws.e-gov.go.jp/document?lawid=425M60080000006
[II–6] NUCLEAR REGULATION AUTHORITY, The Regulatory Guide of NRA Ordinance on Technical Standards for Commercial Power Reactor Facilities, NRA JAPAN, Tokyo (2023) (in Japanese),
https://www.nra.go.jp/data/000382457.pdf
[II–7] NUCLEAR REGULATION AUTHORITY, Guidelines for Design Against Tsunami for Approval of Construction Plan, NRA, Tokyo (2023) (in Japanese),
https://www.nra.go.jp/data/000418541.pdf
[II–8] JAPAN SOCIETY OF CIVIL ENGINEERS, Tsunami Assessment Method for Nuclear Power Plants in Japan 2016, JSCE, Tokyo (2016) (English translation: 2020),
https://committees.jsce.or.jp/ceofnp04/node/11/
[II–9] JAPAN SOCIETY OF CIVIL ENGINEERS, Tsunami Assessment Method for Nuclear Power Plants in Japan, JSCE, Tokyo (2002) (English translation: 2006),
http://committees.jsce.or.jp/ceofnp/system/files/JSCE_Tsunami_060519.pdf
[II–10] PORT AND HARBOURS ASSOCIATION OF JAPAN, Technical Standards and Commentaries for Port and Harbour Facilities in Japan, PHAJ, Tokyo (2018).
[II–11] MINISTRY OF LAND, INFRASTRUCTURE, TRANSPORT AND TOURISM, Guidelines for Tsunami Resistant Design of breakwaters, MLIT, Tokyo (2013) (in Japanese),
http://www.mlit.go.jp/common/001012142.pdf
[II–12] MINISTRY OF LAND, INFRASTRUCTURE, TRANSPORT AND TOURISM, Notification No. 1318, Definition of Safe Structural Systems Against Tsunami to Presume Tsunami Inundation, MLIT, Tokyo (2011).
[II–13] JAPAN ELECTRIC ASSOCIATION, Tsunami Design Engineering Code for Nuclear Power Plants, Rep. JEAC 4629-2021, JEA, Tokyo (2021).
[II–14] JAPAN ELECTRIC ASSOCIATION, Technical Code for Aseismic Design of Nuclear Power Plants, Rep. JEAC 4601-2021, JEA, Tokyo (2021).
[II–15] JAPAN ELECTRIC ASSOCIATION, Technical Guidelines for Aseismic Design of Nuclear Power Plants, Rep. JEAG 4601-2021, JEA, Tokyo (2021).
[II–16] JAPAN ELECTRIC ASSOCIATION, Technical Guidelines for Watertight Facilities, Rep. JEAG 4630-2020, JEA, Tokyo (2020).
[II–17] ATOMIC ENERGY SOCIETY OF JAPAN, Implementation Standard Concerning the Tsunami Probabilistic Risk Assessment of Nuclear Power Plants, Rep. AESJ-SC-RK004:2016, AESJ, Tokyo (2019).

[II–18] ATOMIC ENERGY SOCIETY OF JAPAN, Implementation Standard Concerning the Tsunami Probabilistic Risk Assessment of Nuclear Power Plants, Rep. AESJ-SC-RK004E:2011, AESJ, Tokyo (2013).

ABBREVIATIONS

BDBEE	beyond design basis external event
DBE	design basis earthquake
DEC	design extension condition
NPP	nuclear power plant
NRA	Nuclear Regulation Authority
NRC	United States Nuclear Regulatory Commission
SCE-DS	systems, components, equipment and distribution systems
SSCs	structures, systems and components
TC	Tsunami Category

CONTRIBUTORS TO DRAFTING AND REVIEW

Abe, H.	Japan Nuclear Energy Safety Organization, Japan
Acharya, S.	Atomic Energy Regulatory Board, India
Ahmad, M.	Pakistan Atomic Energy Commission, Pakistan
Anzhar, K.	National Nuclear Energy Agency, Indonesia
Azuma, K.	Japan Nuclear Energy Safety Organization, Japan
Ebisawa, K.	Japan Nuclear Energy Safety Organization, Japan
Ferrante, F.	Nuclear Regulatory Commission, United States of America
Fujii, N.	Tokyo Electric Power Services, Japan
Godoy, A.	Consultant, Argentina
Hardy, G.	Simpson Gumpertz & Heger, United States of America
Huffman, K.	Electric Power Research Institute, United States of America
Hibino, K.	International Atomic Energy Agency
Imamura, F.	Tohoku University, Japan
Iwabuchi, Y.	International Atomic Energy Agency
Javaid, H.	Pakistan Atomic Energy Commission, Pakistan
Johnson, J.J.	James J. Johnson and Associates, United States of America
Kameda, H.	Kyoto University, Japan
Kammerer, A.	Bechtel Corporation, United States of America
Kihara, N.	Central Research Institute of Electric Power Industry, Japan
Kim, M.K.	Korea Atomic Energy Research Institute, Republic of Korea
Li, Z.	China Guangdong Nuclear Power Design Company, China

Matsuyama, M.	Central Research Institute of Electric Power Industry, Japan
Mihara, Y.	Kajima Corporation, Japan
Nagasawa, K.	International Atomic Energy Agency
Park, J.-H.	Korea Atomic Energy Research Institute, Republic of Korea
Pisharady, A.S.	Atomic Energy Regulatory Board, India
Richards, J.M.	Electric Power Research Institute, United States of America
Serbanescu, D.	Consultant, Romania
Sugino, H.	Japan Nuclear Energy Safety Organization, Japan
Varpasuo, P.	Fortum Power and Heat Oy, Finland
Viallet, E.	Electricité de France, France
Videla, M.L.	International Atomic Energy Agency
Yagii, T.	Tokyo Electric Power Company Holdings, Japan
Yalciner, A.C.	Middle East Technical University, Türkiye
Yamazaki, H.	Japan Nuclear Energy Safety Organization, Japan
Yamazaki, T.	Japan Nuclear Safety Institute, Japan
Yi, S.	KEPCO Engineering & Construction, Republic of Korea
Yoshimura, S.	University of Tokyo, Japan

Consultants Meetings

Tokyo, Japan: 27–31 May 2013

Vienna, Austria: 25–26 July 2013, 29 July–2 August 2013,
18–22 November 2013, 26–27 May 2014

Washington, DC, United States of America: 23–24 September 2013

San Francisco, California, United States of America: 12–13 February 2014

Kyoto, Japan: 31 March–4 April 2014

IAEA
International Atomic Energy Agency

CONTACT IAEA PUBLISHING

Feedback on IAEA publications may be given via the on-line form available at:
www.iaea.org/publications/feedback

This form may also be used to report safety issues or environmental queries concerning IAEA publications.

Alternatively, contact IAEA Publishing:

Publishing Section
International Atomic Energy Agency
Vienna International Centre, PO Box 100, 1400 Vienna, Austria
Telephone: +43 1 2600 22529 or 22530
Email: sales.publications@iaea.org
www.iaea.org/publications

Priced and unpriced IAEA publications may be ordered directly from the IAEA.

ORDERING LOCALLY

Priced IAEA publications may be purchased from regional distributors and from major local booksellers.

25-00153E